DESIGN BRIEFS
STUDENT'S RESEARCH BOOK

ROY LEE
Horringer Court Middle School, Bury St Edmunds, Suffolk

JOHN ALDRIDGE
Horringer Court Middle School, Bury St Edmunds, Suffolk

CAMBRIDGE
UNIVERSITY PRESS

For Carol and Marian

Published by the Press Syndicate of the University of Cambridge
The Pitt Building, Trumpington Street, Cambridge CB2 1RP
40 West 20th Street, New York, NY 10011–4211, USA
10 Stamford Road, Oakleigh, Melbourne 3166, Australia

© Cambridge University Press 1989

First published 1989
Third printing 1994

Printed in Great Britain by GreenShires Print Ltd, Kettering, Northants.

Lee, Roy
Design briefs: student's research book.
1. Engineering Design, – For schools
I. Title II. Aldridge, John
620′.00425

ISBN 0 521 34827 7

VN

The publishers would like to thank the following for permission to reproduce their photos.
Structures: Ironbridge Gorge Museum Trust, Freeman Fox Ltd (Humber Bridge), Costain Group plc (building site), J. Smith & Sons (Clerkenwell) Ltd (metals). *Mechanisms:* National Railway Museum, York. *Energy & control:* All-Sport (sailing), Wind energy group, Hubert Herr (clock), Boxmag Rapid (electromagnet), Aeromodeller. *Using science:* All-Sport (skater, Fomula one), London Transport (bus), Royal National Lifeboat Institution (dinghy), Barnaby's (steel band). *Using materials:* Welding Institute, Addis (plastics), Ancasta (boat), Timber Research and Development Association, Henkel Home Improvements and Adhesive products.

Other photographs taken by Nick Brown, Charlotte Attwood and Nigel Luckhurst.

Cover design by Chris McLeod, photographs by Nigel Luckhurst.

CONTENTS

INTRODUCTION

USING THIS BOOK

At the heart of the subject of Craft, Design and Technology is an exciting process of practical problem-solving. Design Briefs will challenge you to design and make your own solutions to design problems. You will sometimes work alone but, as in industry, you will often be a member of a team. Your teacher will be there to guide you but most of the design decisions will be yours. The purpose of this book is to provide you with a databank of technical and scientific information.

The information is arranged in six sections and there is an index and glossary at the back of the book. The index gives you the section and sheet number where you can expect to find the information you need.

When you come across words printed in **bold**, you will know you can find a definition of the word in the glossary. The glossary is also useful for revision purposes. Try checking the glossary from time to time to see how your knowledge of technical and scientific terms is improving.

WORKING WITH DESIGN BRIEFS

Read carefully through your Brief and decide what further information you need in order to start making decisions. The Information Sheets and photographs will provide you with lots of ideas, but try to look further afield too. Useful sources include books, magazines and photographs in libraries but, in addition, do not ignore television, newspapers, your home, your neighbourhood, local shops and industry. Designers use ideas from every part of their environment. Keep your eyes open and try to train yourself to do the same.

Share the information you have found with other people in your class and then explore a range of solutions as a class or in groups. This process is called brain-storming and the aim is to think of as many solutions as possible. Don't worry if lots of the ideas probably wouldn't work. You can worry about practical things later. Some of the greatest technical advances have come from the craziest ideas. Many people laughed at Cockcroft, the inventor of the hovercraft, when he suggested a vehicle could be made to float on a cushion of air!

After the brain-storming session, it is time to set your feet back on the ground and to look at the practical difficulties of each of the proposed solutions. On your own or in a group carefully consider all the limitations, including your time, your skill, money and materials. Make a short list of possible solutions. Try to include at least two, and up to five, ideas. In selecting your chosen solution, it is usually helpful to make accurate drawings or even a scale **mock-up**.

Having chosen your design, you are now ready to work out the details of how you intend to make it. Will you work alone or with others? What tools and materials do you need? Do you need any further information? How long will it take? Don't forget to check your actual progress against your plan from time to time. Don't be afraid to make changes if you think them necessary.

When you have completed your project, we come to the vital stage of **evaluation**. Your teacher will give you an evaluation sheet to help you. There are many things to consider. To what extent have things worked out as you planned? Did you choose the right materials? Was your timetable realistic? Are you satisfied with the appearance of your product? Finally, does your product match the Brief you were given? The fitness of your product for the purpose it is intended is the most important consideration of all.

STRUCTURES

Clockwise from top left. The Iron Bridge, Coalbrookdale; a building site; metal tubes; the Humber Bridge; the Schlumberger building, Cambridge. **How many different kinds of structures can you find?**

TENSION – COMPRESSION   SHEARING – TORSION   TIES AND STRUTS

STRUCTURE

Structure is the name given to a supporting framework. A structure must be able to resist forces acting on it, without collapsing. The forces acting on a structure are either static (not moving) or dynamic (moving). Dynamic forces are much more destructive than static ones.

FORCES

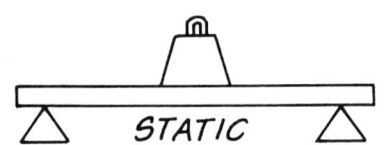 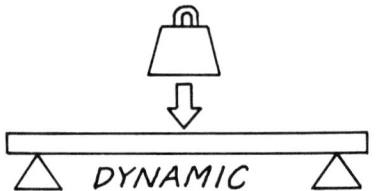

STATIC   DYNAMIC

There are several different ways that forces can act on a structure – often at the same time.

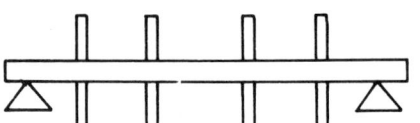

COMPRESSION (CLOSER TOGETHER)

TENSION (FURTHER APART)

TENSION AND COMPRESSION

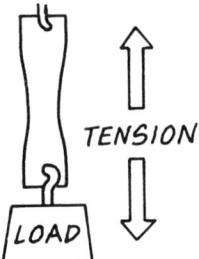

TENSION

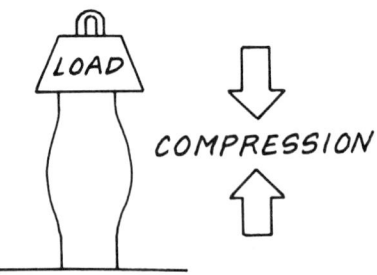

COMPRESSION

Tension is a force which is trying to stretch the material.
Compression is a force which is trying to squash the material.

SHEARING

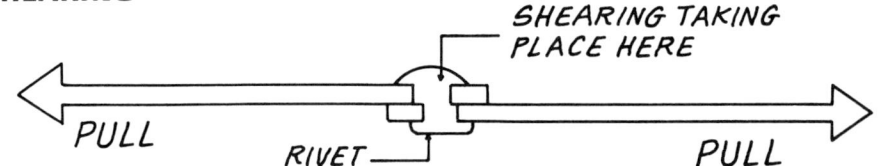

SHEARING TAKING PLACE HERE

PULL   RIVET   PULL

Shearing is a force which is trying to move one part of the material in the opposite direction to the other.

TORSION

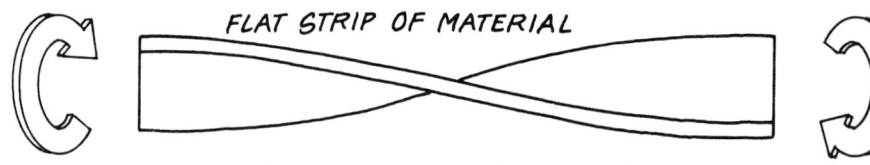

FLAT STRIP OF MATERIAL

Torsion is a force which is trying to twist the material.

TIES AND STRUTS

In structures, parts that are in tension are called ties and are usually thinner than the parts that are under compression called struts.

LADDERS

STRUTS

TIES

FLAGPOLE

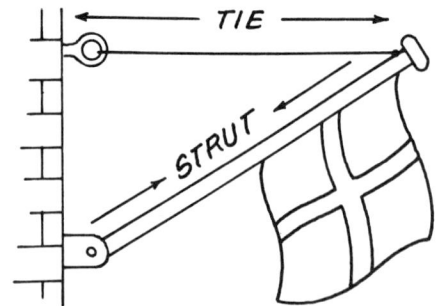

TIE

STRUT

FOLDING

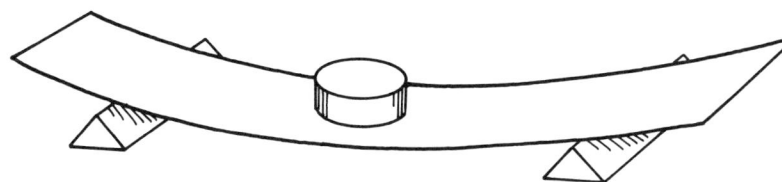

Most thin sheet materials are **flexible**. If you fold the sheet in half and double its thickness it will still be flexible – in other words you will not have increased its **rigidity** by very much!

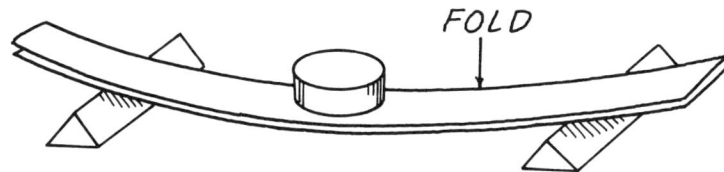

Open out the fold until each half is at right-angles to the other.

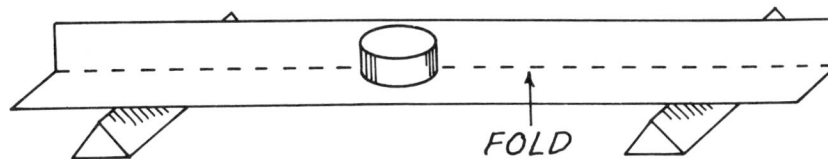

You should now find that your sheet is much more rigid. This is a simple L-shaped girder.

GIRDERS

Here are some more complex girders using more folds.

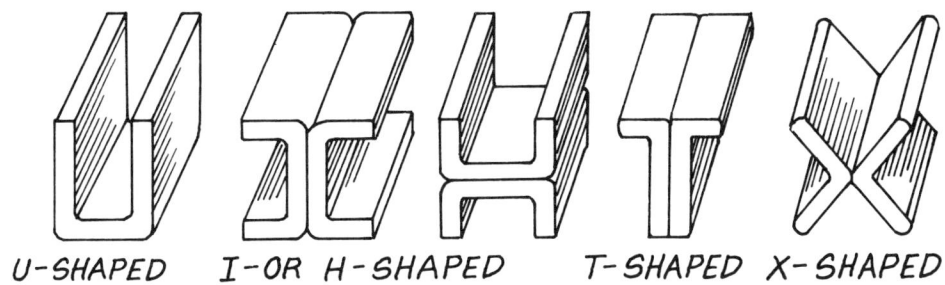

U-SHAPED   I- OR H-SHAPED   T-SHAPED   X-SHAPED

See if you can work out some of your own.

BOX GIRDERS

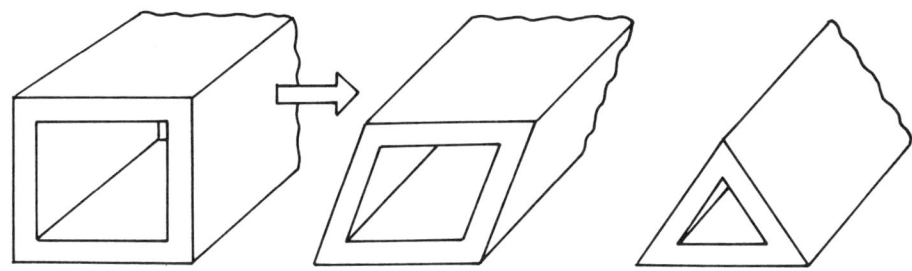

Box girders are ones which form an enclosed space. The square section can be squashed to one side whilst the triangular section is still rigid. The tube is also a kind of box girder – look at a bicycle frame.

3

CORRUGATION   CELLULAR FORMS

CORRUGATION

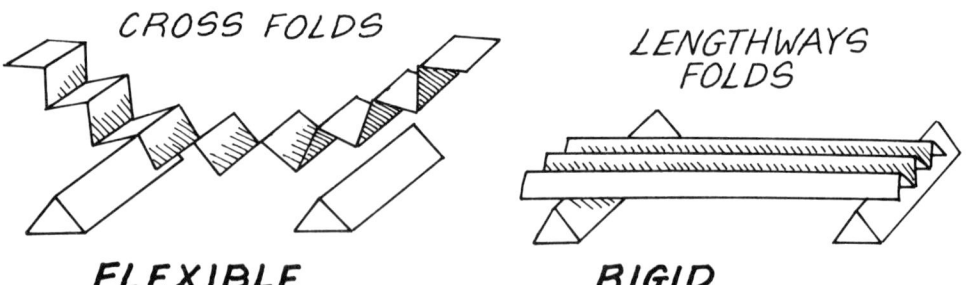

CROSS FOLDS   LENGTHWAYS FOLDS

FLEXIBLE   **RIGID**

Putting **corrugations** (repeated or concertina folds) into sheet material is another way of making it more rigid. Notice that the increased **rigidity** is only in the direction at right-angles to the folds.

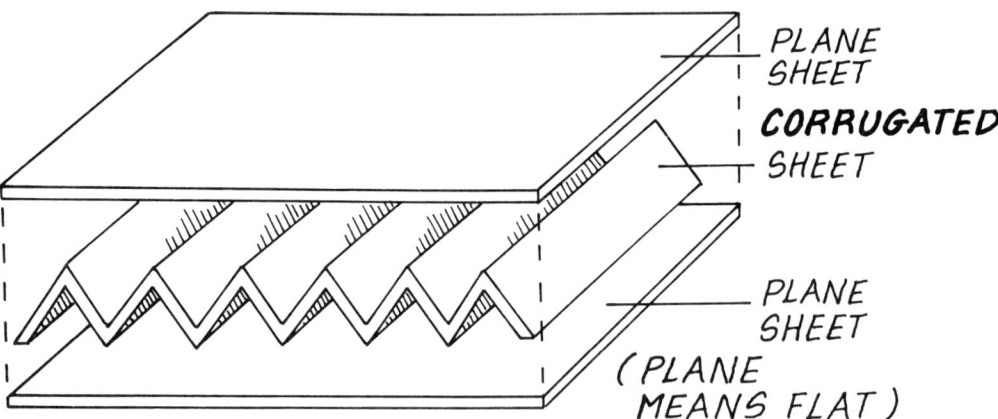

PLANE SHEET

CORRUGATED SHEET

PLANE SHEET

(PLANE MEANS FLAT)

If a corrugated sheet is glued (or welded) between two plane sheets, some of the flexibility is reduced and the 'sandwich' produced is more rigid in all directions. (Note that it is still most rigid at right-angles to the direction of the folds.) Look at corrugated card and iron.

CELLULAR FORMS

Some of the shapes often used for **cellular** structures.

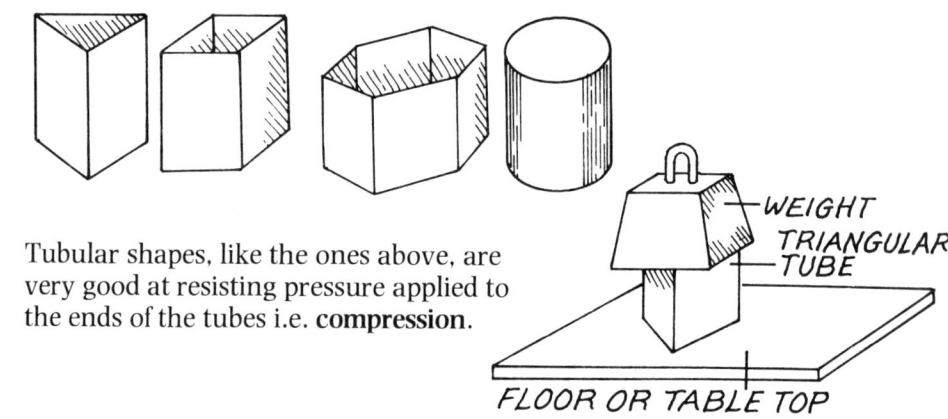

WEIGHT
TRIANGULAR TUBE

FLOOR OR TABLE TOP

Tubular shapes, like the ones above, are very good at resisting pressure applied to the ends of the tubes i.e. **compression**.

This has been made use of where strength and lightness are both extremely important, e.g. the floor panels of an aeroplane. Cellular structures can also save money, because a cheap material can be used in the place of a dear one. Look at the inside of a hardboard-faced door if you can.

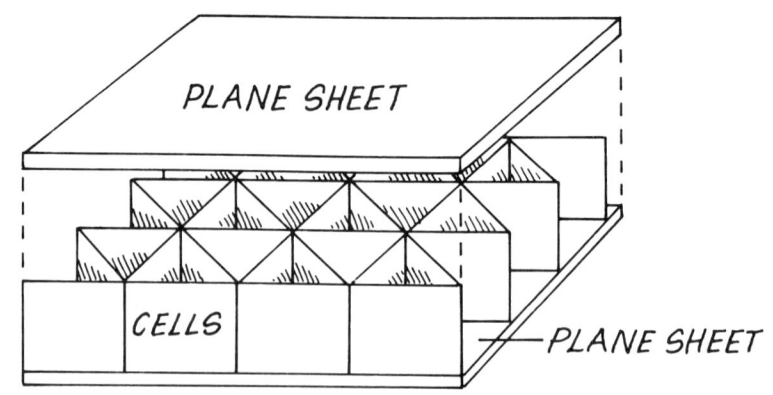

PLANE SHEET

CELLS

PLANE SHEET

TRIANGULATION

So far we have looked at ways of making our beams and sheet materials more **rigid**. Now we will look at ways of making the structures themselves more rigid.

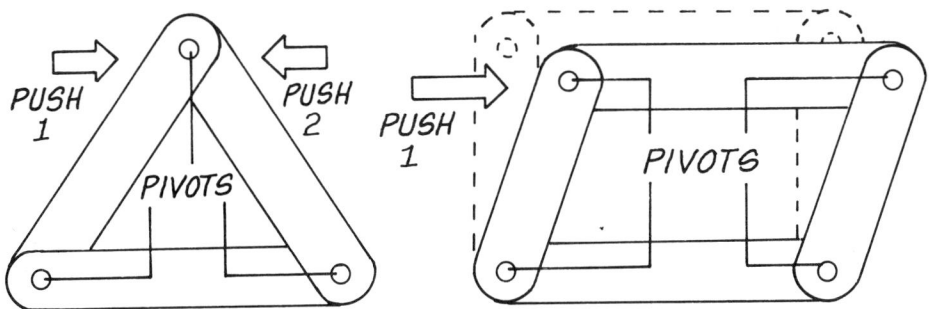

If we make these structures with strips of stiff card and paper rivets, we discover that the triangular shape is rigid and the rectangular shape is not. Can we make the rectangle rigid?

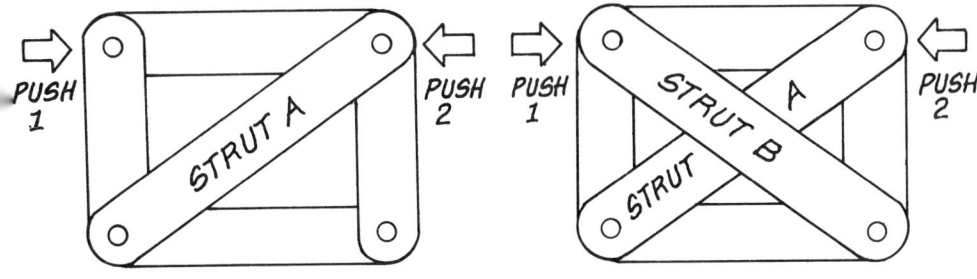

If we add a further strip (**A**) diagonally, our rectangle becomes rigid. Can you see why this is? The rectangle has been divided into two triangles. If we add yet another strip (**B**), the rectangle does not become any more rigid than with only (**A**). Strip (**B**) is what we call redundant, i.e. not needed.

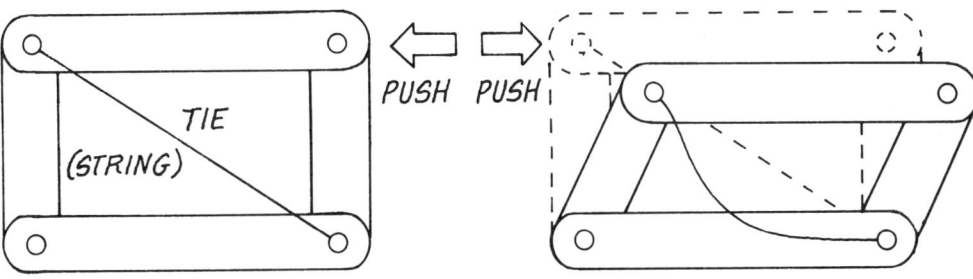

If we use a tie instead of a strut then the rectangle is rigid in one direction but not the other.

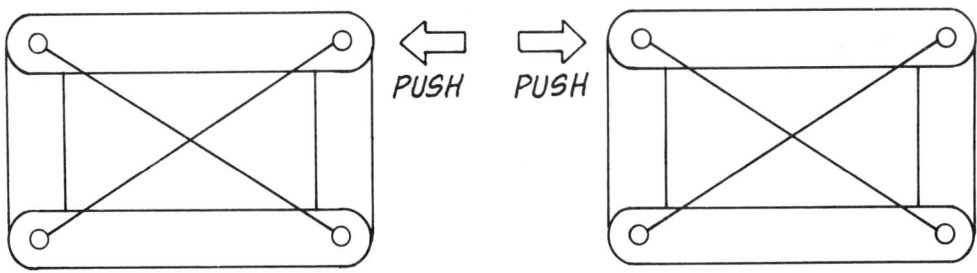

By adding the second tie as a diagonal, the rectangle now becomes totally rigid again. Obviously ties are much lighter than struts, so where weight is important then ties are an attractive alternative to struts even though two are needed. (Look at the bracing wires on early biplanes.) The main problem with ties is getting the tension right.

With care, many rigid, stable and lightweight structures can be built using the principles of triangulation. (Look at cranes, bridges and bicycle frames.)

USING SHEET MATERIALS

USING SHEET MATERIALS

Another way of making a rectangular frame more rigid is by fixing a sheet material to the frame, in this case we will use cardboard.

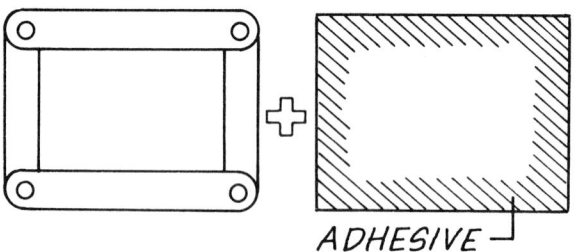 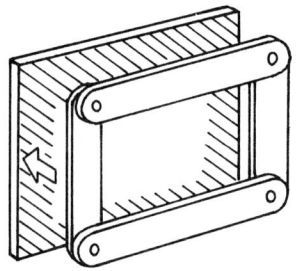

ADHESIVE

Pushing against the corners will show the frame to be far more rigid than the unsupported frame. We can get a similar increased rigidity if we take a rectangular sheet and fold up the edges all the way round to make a tray.

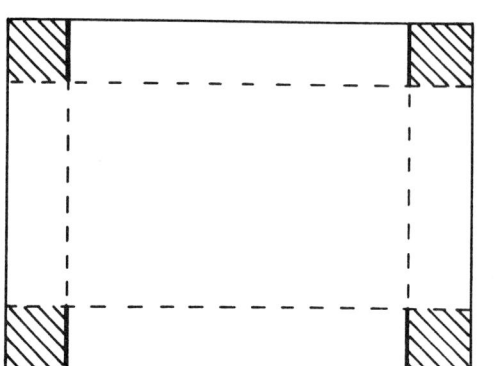

DOTTED LINE = FOLD
SOLID LINE = CUT
⬚ GLUE TAB

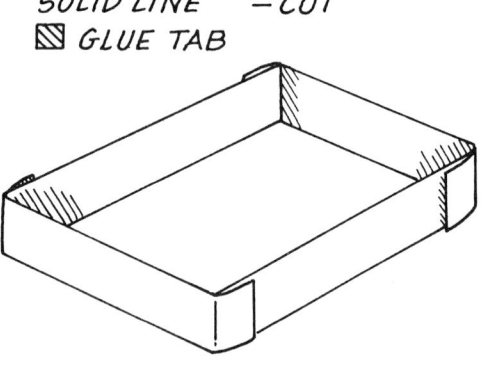

The flat shape on the left is called a 'net'.

This method will also make other shapes more rigid.

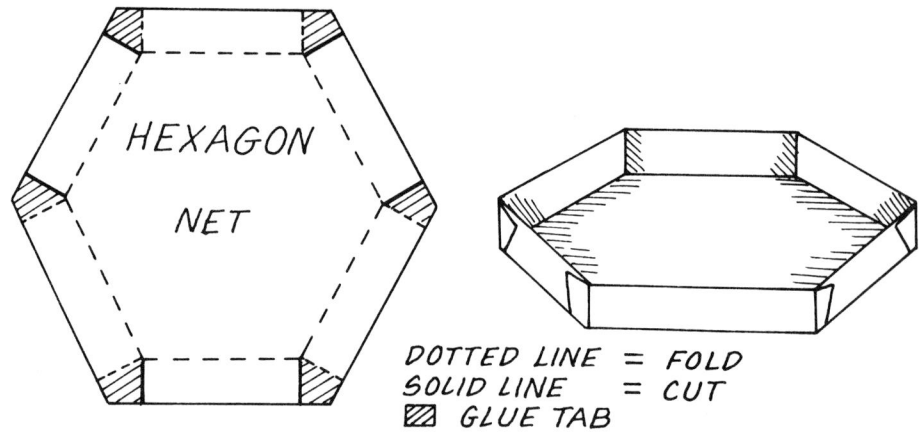

HEXAGON
NET

DOTTED LINE = FOLD
SOLID LINE = CUT
⬚ GLUE TAB

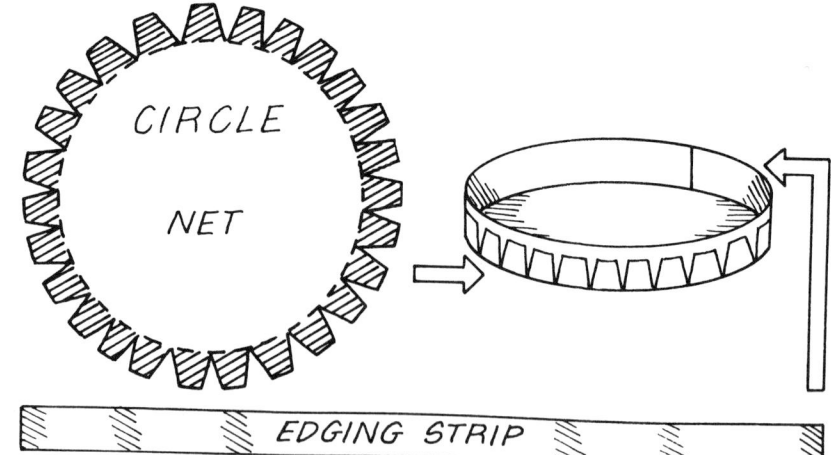

CIRCLE
NET

EDGING STRIP

This idea works with all sheet materials, although the method of folding up the edge will depend on the material used. (Look at a pressed metal drinks tray, a wooden drawer, a plastic drawer and a yoghurt carton.)

CUBE

Here are useful nets for a cube and cuboid. They could be used for packaging or as the basis of papier mâché bodyshells.

CUBOID

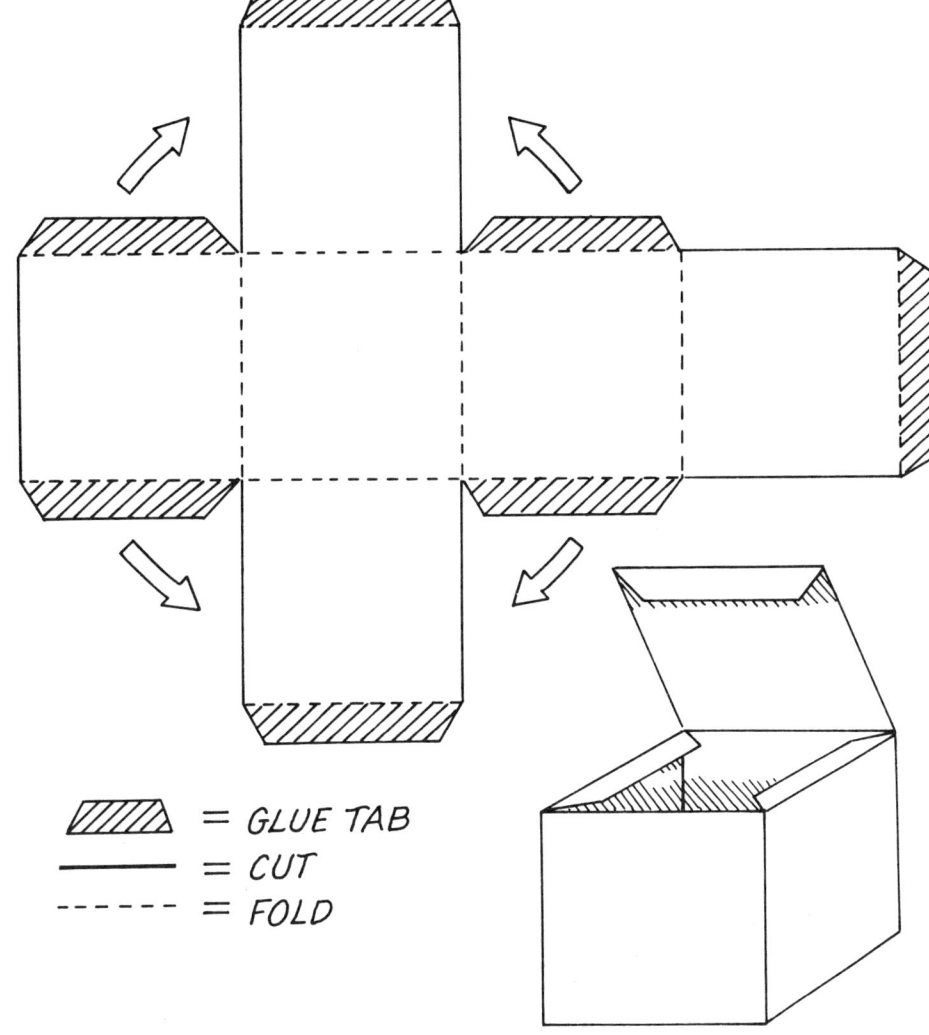

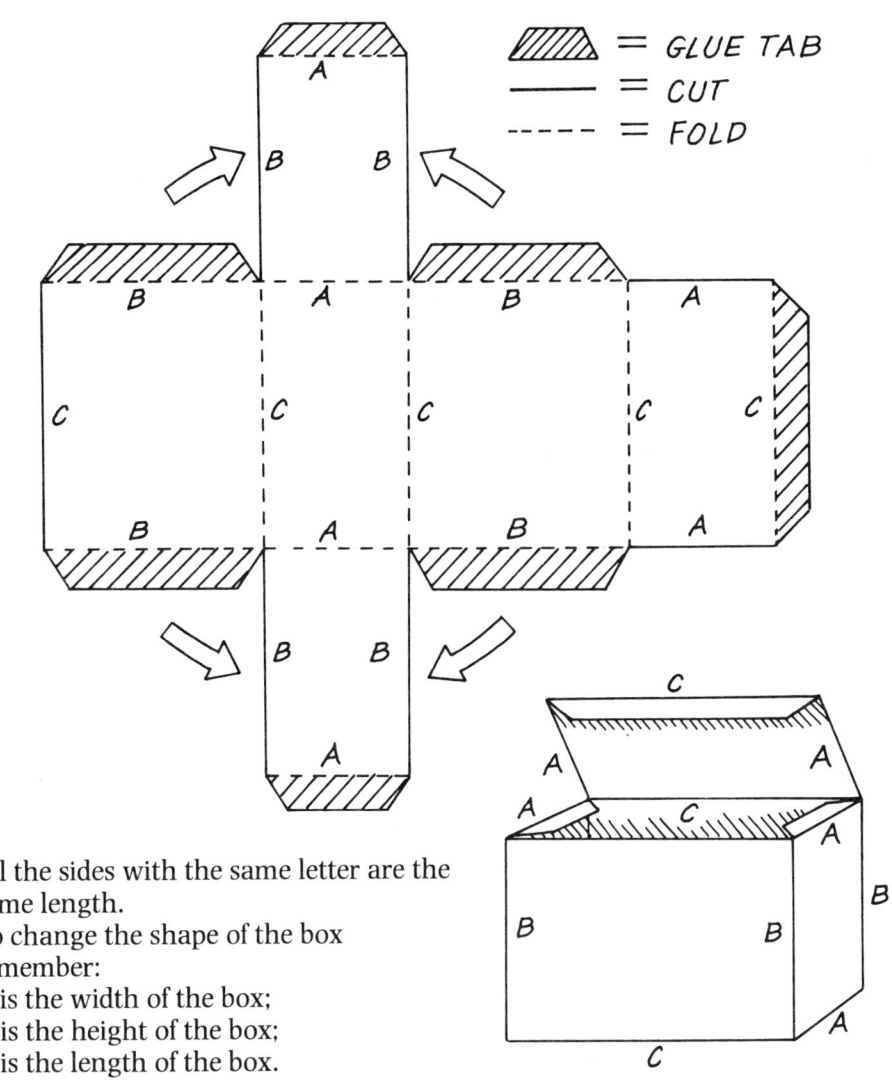

= GLUE TAB
= CUT
= FOLD

All the sides with the same letter are the same length.
To change the shape of the box remember:
A is the width of the box;
B is the height of the box;
C is the length of the box.

7

USEFUL NETS   TRIANGULAR PRISM

Here are useful nets for triangular prisms. They could be used for packaging or as the basis of papier mâché bodyshells.

TRIANGULAR PRISM (REGULAR)

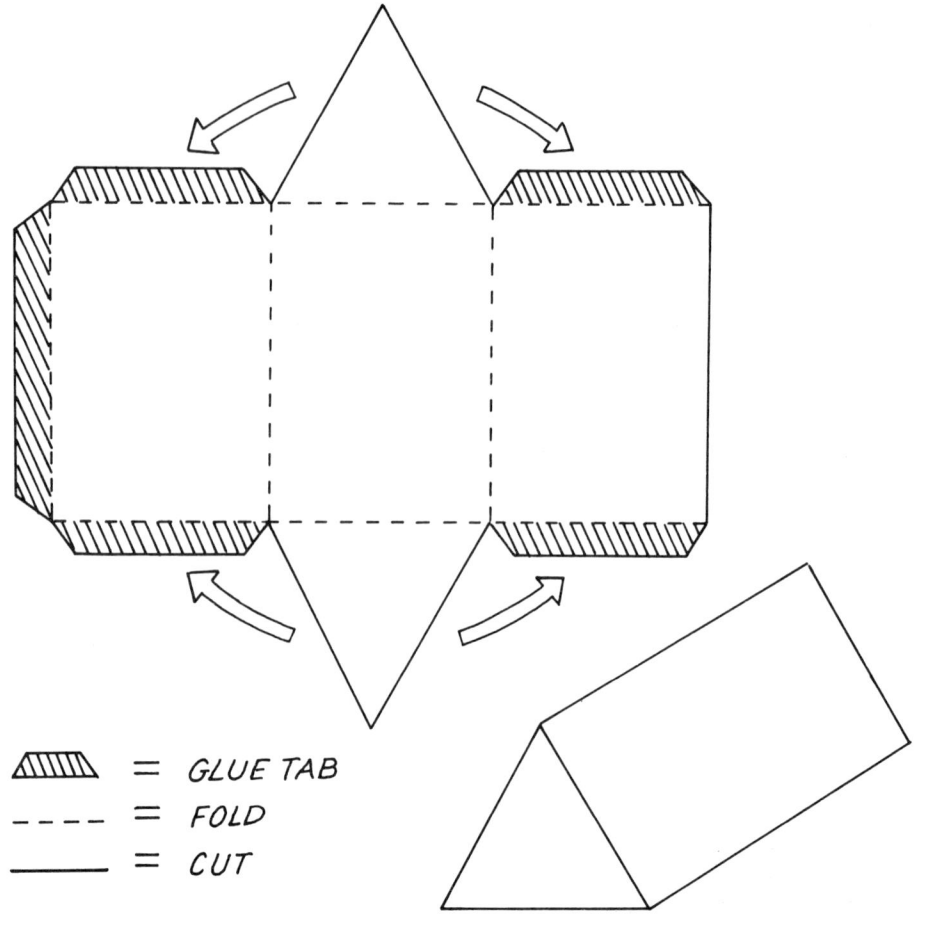

= GLUE TAB

- - - - = FOLD

———— = CUT

TRIANGULAR PRISM (IRREGULAR)

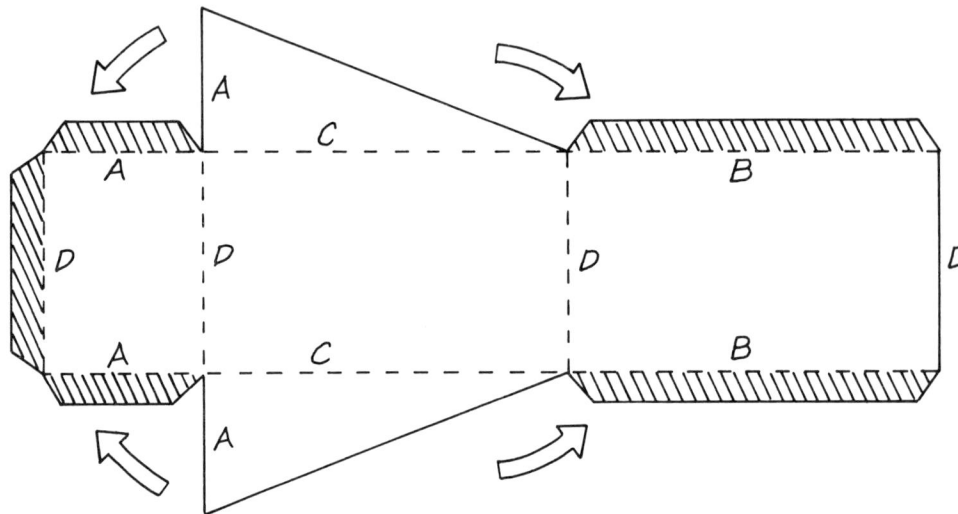

= GLUE TAB

- - - - = FOLD

———— = CUT

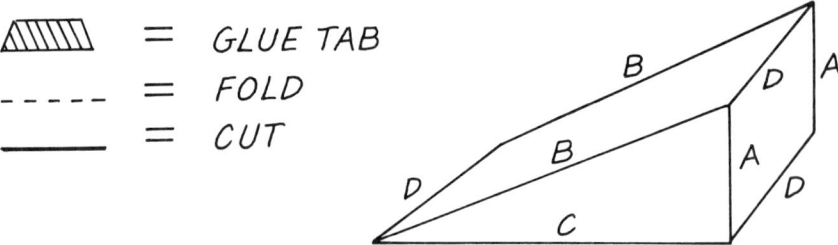

All the sides with the same letters are the same length. Any change to **A** or **B** or **C** could change at least one, if not both, of the remaining sides of the triangle.

Changing **A** and **B** alters the height of the prism.

Changing **D** alters the width of the prism.

Changing **C** (together with **A** or **B**) alters the length.

STRUCTURES

USEFUL NETS   PYRAMID

Here are useful nets for pyramids. They could be used for packaging or as the basis of papier mâché bodyshells.

PYRAMID (REGULAR)

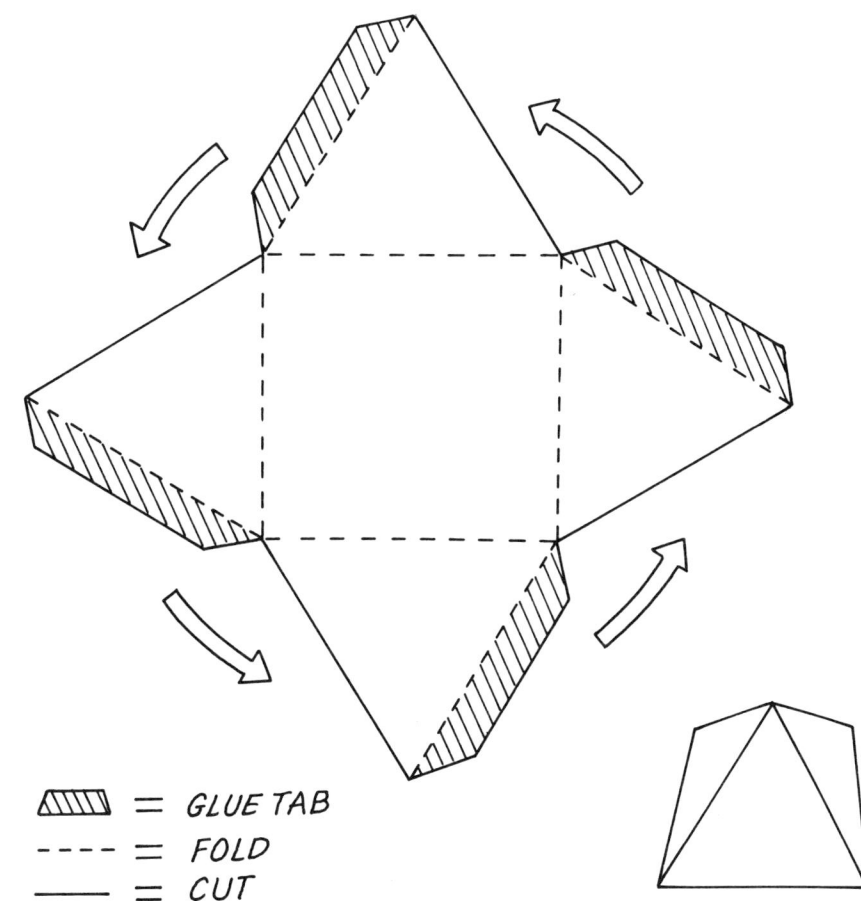

```
///// = GLUE TAB
- - - = FOLD
____ = CUT
```

PYRAMID (IRREGULAR)

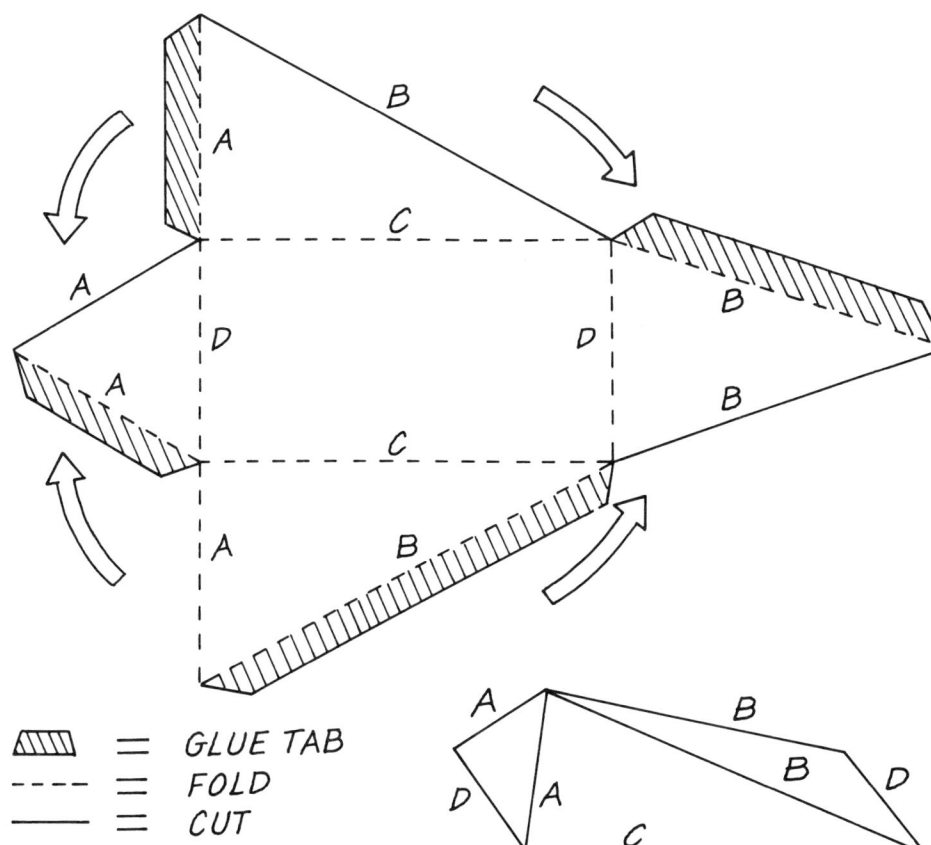

```
///// = GLUE TAB
- - - = FOLD
____ = CUT
```

To change the shape of the pyramid remember:
any change to **A** or **B** or **C** could change at least one if not both of the remaining sides of the triangle.
Changing **A** and **B** alters the height of the pyramid.
Changing **D** alters the width (and also affects height).
Changing **C** alters the length (and also affects height).

MECHANISMS

Clockwise from top left. Belt drive used on a pillar drill machine; a steam locomotive showing drive levers and cranks; lock mechanism (ratchet and pawl); block and tackle pulley; lock mechanism (rack and pinion), note the cranking handle. **Which of these mechanisms do you think is the oldest? Why?**

LEVERS

A lever is a rigid bar or beam which can turn freely round a fixed point (**pivot**). (The fixed point is also called a **fulcrum**.)

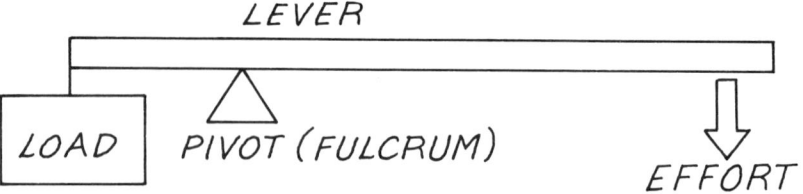

If you have ever opened a tin using a screwdriver or scissors to prise off the lid, then you have used a lever. In this case by using a lever a smaller effort is needed to open the tin than trying to pull off the lid with your fingers. However, the hand pressing the screwdriver handle moves much further than the tin lid so that a small effort moving a large distance moves a large load a small distance.

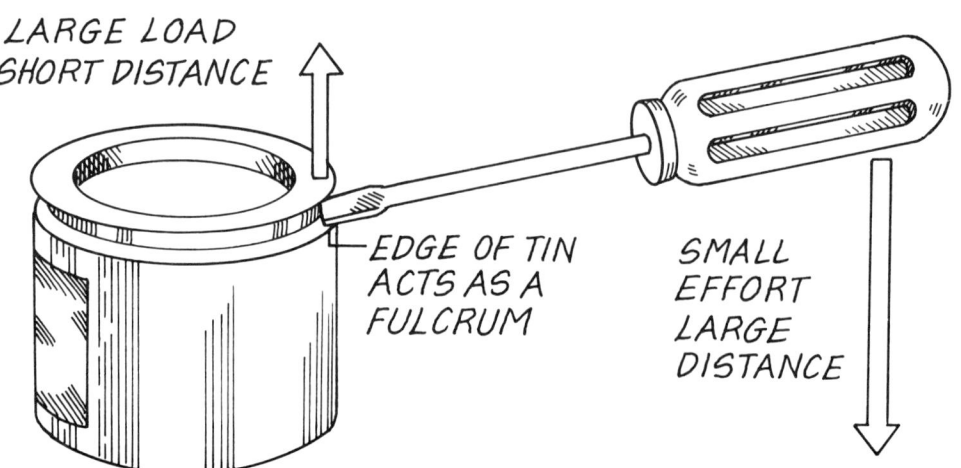

Sometimes the opposite effect is put to use. Here is a weighing device called a Roman Steelyard.

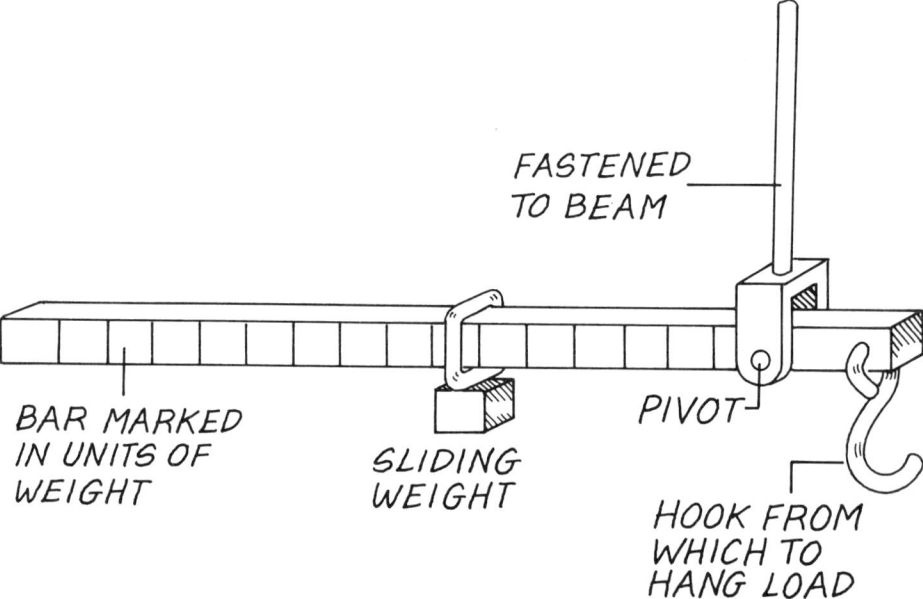

It is used for weighing heavy objects. As you can see, a short movement of the load causes a large movement of the long arm. The weight of the load on the hook is found by sliding the moveable weight along the bar until the bar is level (balance) and reading off the number of divisions between the pivot and the sliding weight. Using the **properties** of the lever in this way increases the sensitivity and therefore the accuracy of the **device**.
(See Mechanisms 2 and Using Science 1 for further information.)

CLASSES OF LEVER

FIRST CLASS

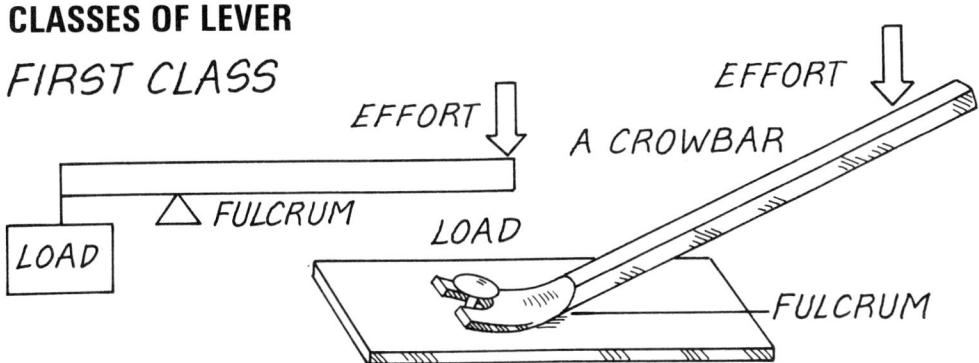

A first class lever has the **fulcrum** between the load and the effort.

SECOND CLASS

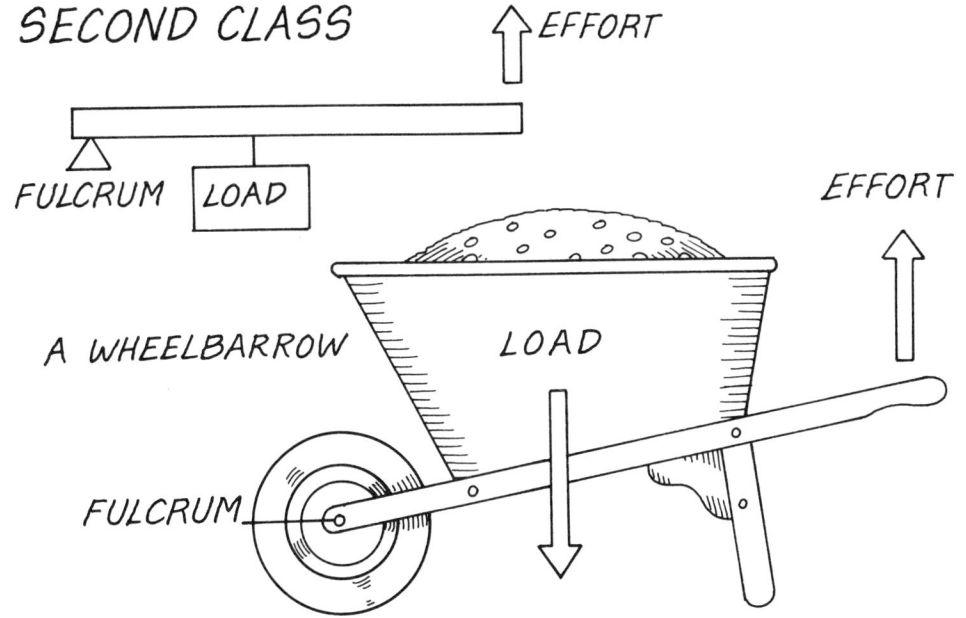

A second class lever has the load between the effort and the fulcrum.

THIRD CLASS

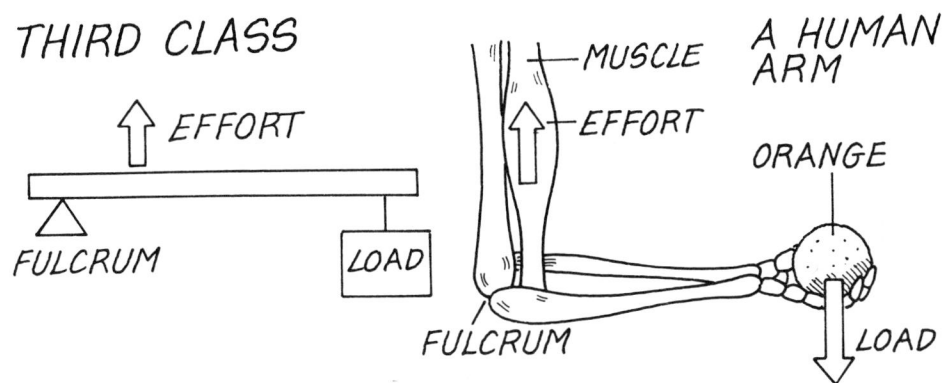

A third class lever has the effort between the fulcrum and the load.

MECHANICAL ADVANTAGE

This is the name given to the ratio of effort to load and is a measure of the efficiency of the lever.

$$\text{Mechanical advantage} = \frac{\text{load}}{\text{effort}}$$

VELOCITY RATIO

This is the name given to the ratio of the distance moved by the effort to the distance moved by the load.

$$\text{Velocity ratio} = \frac{\text{distance moved by effort}}{\text{distance moved by load}}$$

SINGLE ARM SYSTEM

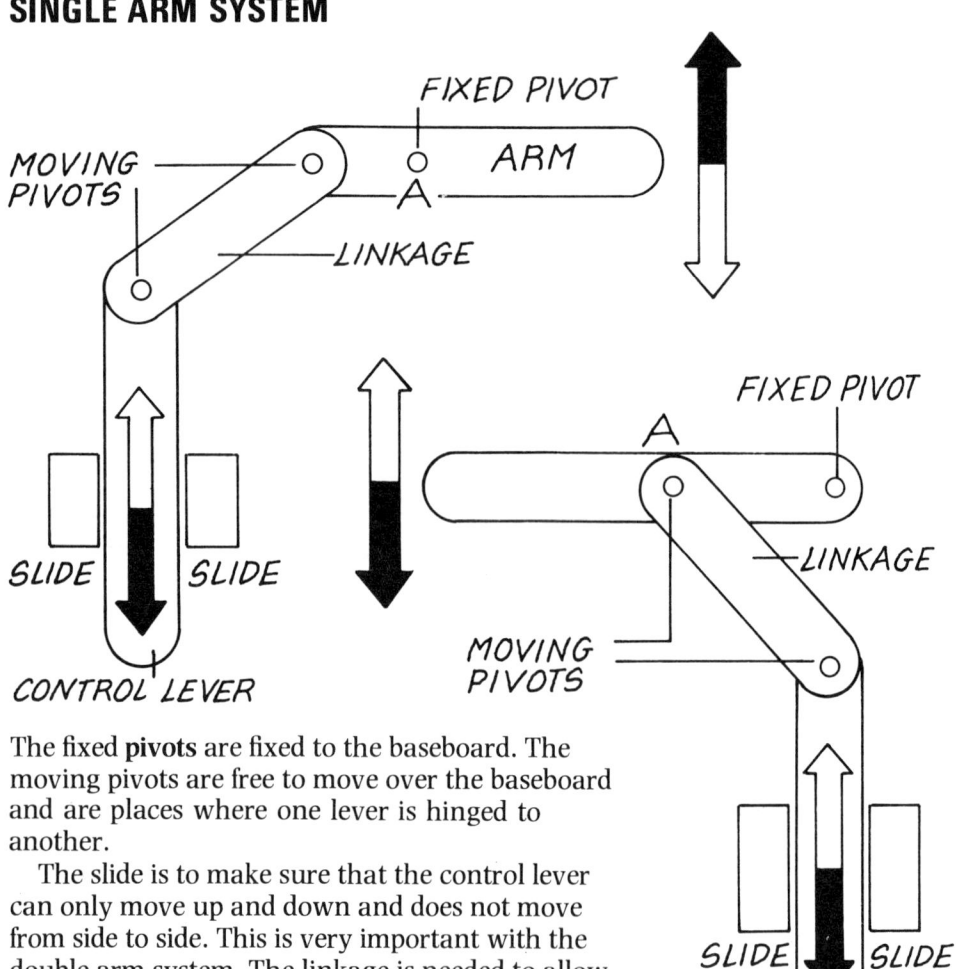

The fixed **pivots** are fixed to the baseboard. The moving pivots are free to move over the baseboard and are places where one lever is hinged to another.

The slide is to make sure that the control lever can only move up and down and does not move from side to side. This is very important with the double arm system. The linkage is needed to allow the up-and-down movement of the control lever to fit the curving movement made by the arm round the fixed pivot. Try moving the pivot marked **A** along the arm. How does this change the movement?

DOUBLE ARM SYSTEM

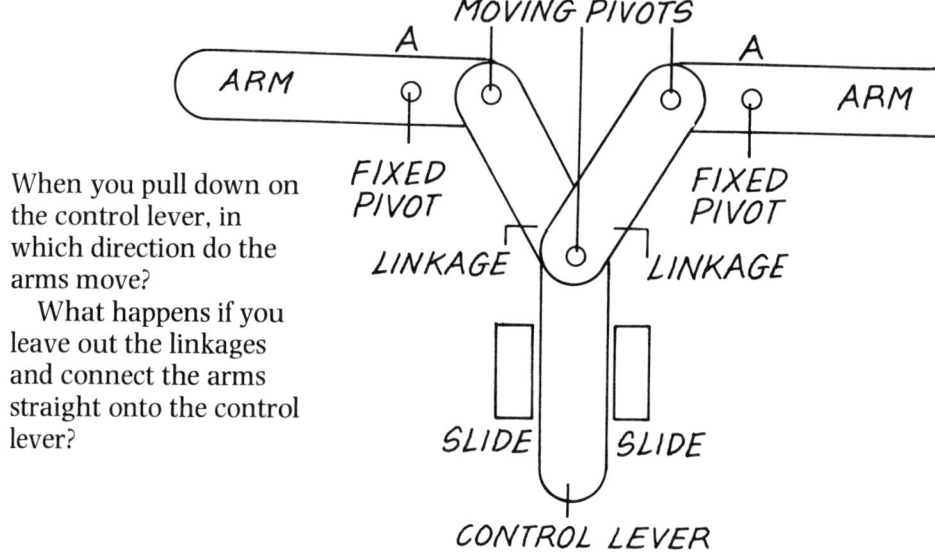

When you pull down on the control lever, in which direction do the arms move?

What happens if you leave out the linkages and connect the arms straight onto the control lever?

How are these two systems different?

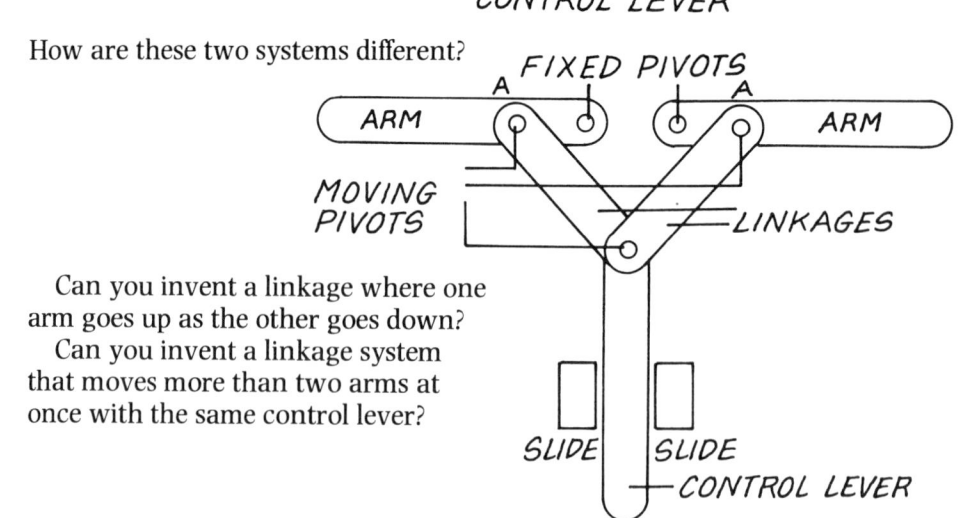

Can you invent a linkage where one arm goes up as the other goes down?

Can you invent a linkage system that moves more than two arms at once with the same control lever?

LINKAGES TO CHANGE DIRECTION OR SIZE OF MOVEMENT

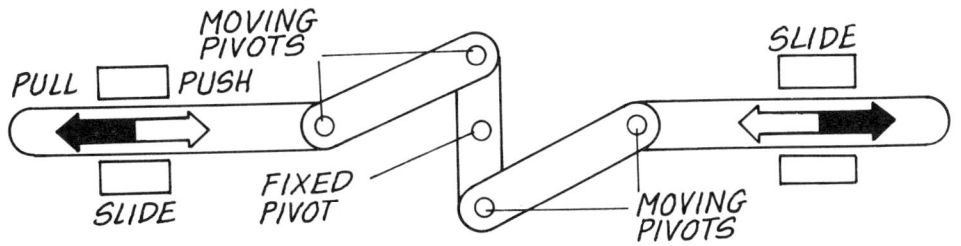

This mechanism could be used for locking a door! Can you see how this might be done? Try putting the fixed pivot nearer one end than the other as shown below.

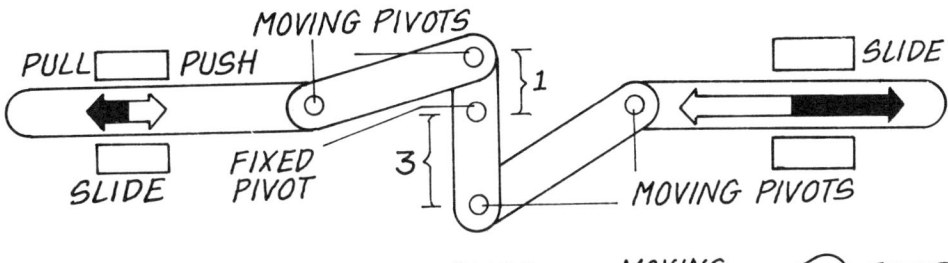

By doing this you can make one of the levers move further than the other. In the example shown above the lever on the right will move three times as far as the lever on the left. In the next drawing on the right we use an L-shaped lever called a Bell Crank. It changes the direction of the movement.

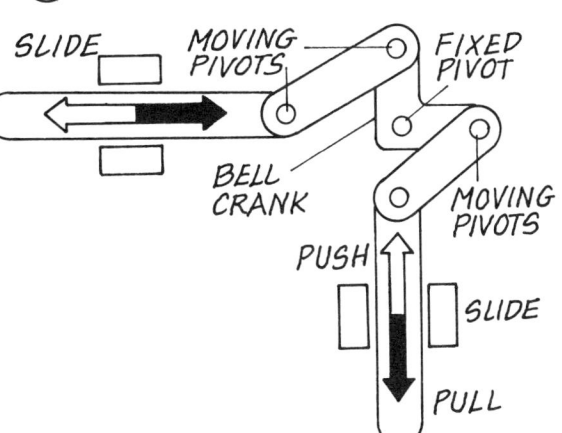

By making the length of one arm of the Bell Crank different from the other, you can make the levers going through the slides move different amounts. In other words you can magnify or reduce movement. Remember there is a price to pay for magnifying the movement. (See Velocity Ratio and Mechanical Advantage.)

MORE CRANKS

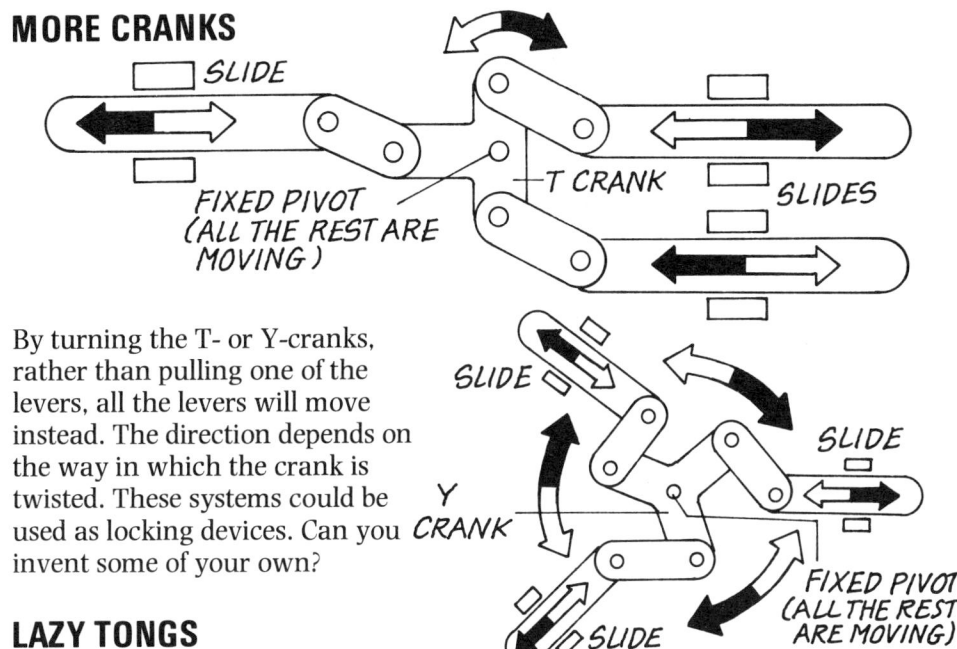

By turning the T- or Y-cranks, rather than pulling one of the levers, all the levers will move instead. The direction depends on the way in which the crank is twisted. These systems could be used as locking devices. Can you invent some of your own?

LAZY TONGS

Here we have a slightly different use of levers. When **A** and **B** are pushed towards each other the tongs get longer. Pulling them apart makes the tongs shorter. How could you make use of this?

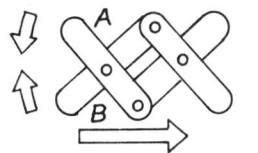
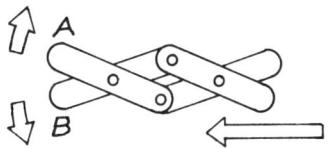

MAKING PULLEYS FROM PLY

Although there are pulleys available ready-made, you might like to have a go at making your own.

Use three discs of three ply. Find the exact centre of each and drill so that each disc is a tight push-fit on the axle. Glue the three discs together.

So that your pulleys run smoothly, it is important that they have good **bearings**. Bearings are the holes in which the axles rotate. They are usually made of metal to reduce friction.

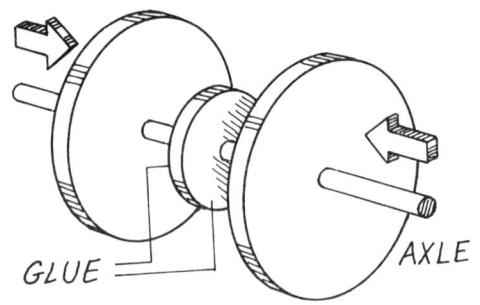

GLUE   AXLE

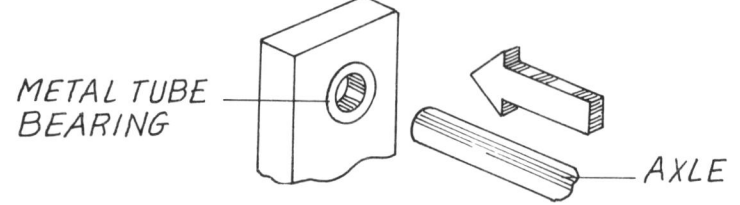

METAL TUBE BEARING   AXLE

CHANGING THE SPEED OF ROTATION

Diameter of middle disc **A** = 2 cm.
Diameter of middle disc **B** = 6 cm.
NOTE Ignore the outer discs in your calculations. The middle disc is the bit the belt is wrapped around.
Divide the diameter of disc **B** by the diameter of disc **A**. The result is the number of turns **A** makes for one turn of **B**.

Example
For every turn B makes, A will turn

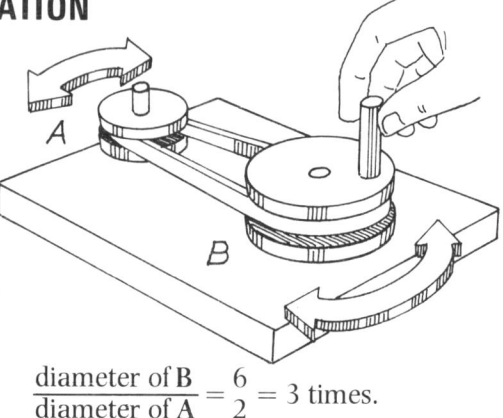

$$\frac{\text{diameter of B}}{\text{diameter of A}} = \frac{6}{2} = 3 \text{ times.}$$

POWER AND PULLEYS

If pulley **A** is the drive **pulley** then **B** will rotate more quickly but with less power, i.e. it stops more easily if a braking force is applied. If pulley **B** is the drive pulley then **A** will rotate more slowly but with greater power.

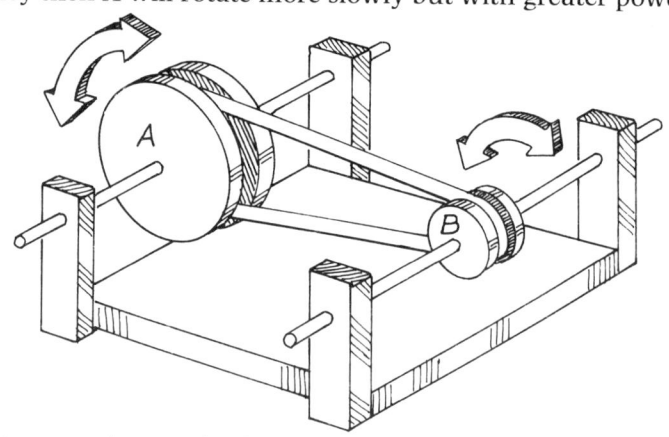

You can increase the speed of **B** by reducing the diameter of **B**. You can reduce the speed of **B** by reducing the diameter of **A**. How would these changes affect the power?

IMPORTANT

The pulleys must be in line with each other.
The axles must be **parallel**.
The belt must be correctly tensioned.

LIFTING PULLEYS

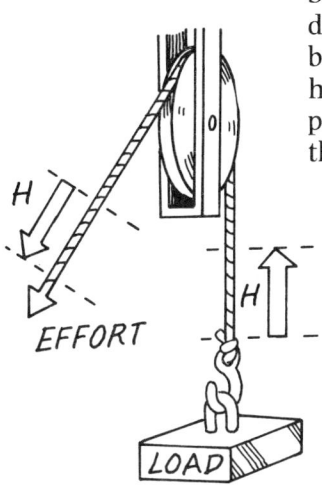

Single pulleys like this are often used to change the direction of the effort. This allows you to use your body weight to help lift heavy weights. It can also help you lift them to a greater height. A single pulley does not allow you to lift a load greater than the effort. To do this you need more pulleys.

BLOCK AND TACKLE

This is the name given to two-part pulley systems. In a two-part pulley system, the weight of the load is shared half each to the rope on either side of the bottom pulley. The tension is spread through the length of the rope so the effort required to lift the load is only half the load. To lift the load through a height (**H**) the effort has to move twice as far (**H** + **H** = 2**H**). If you have studied levers (Mechanisms 2) this should sound familiar!

$$\text{Mechanical advantage} = \frac{\text{load}}{\text{effort}} = 2$$

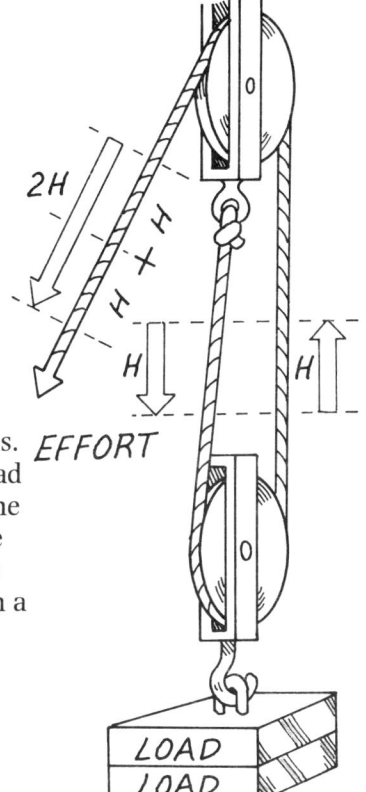

SAME EFFORT LIFTS
TWICE THE LOAD

$$\text{Velocity ratio} = \frac{\text{distance moved by effort}}{\text{distance moved by load}} = 2$$

With a three-pulley system the load is shared between three ropes. This time the effort required to lift the load is only a third of the load.

$$\text{Mechanical advantage} = \frac{\text{load}}{\text{effort}} = 3$$

The effort has to move three times as far (**H** + **H** + **H** = 3**H**) to lift the load a distance of (**H**).

$$\text{Velocity ratio} = \frac{\text{distance moved by effort}}{\text{distance moved by load}} = 3$$

You may have noticed that in each case the mechanical advantage–velocity ratio of a system is the same as the number of pulleys in that system (and the pulleys are all the same size).

We have ignored the effect of **friction**, but in a system of more than six pulleys it starts making a noticeable difference. (Plus of course the weight of the pulleys and ropes begins to make a difference.)

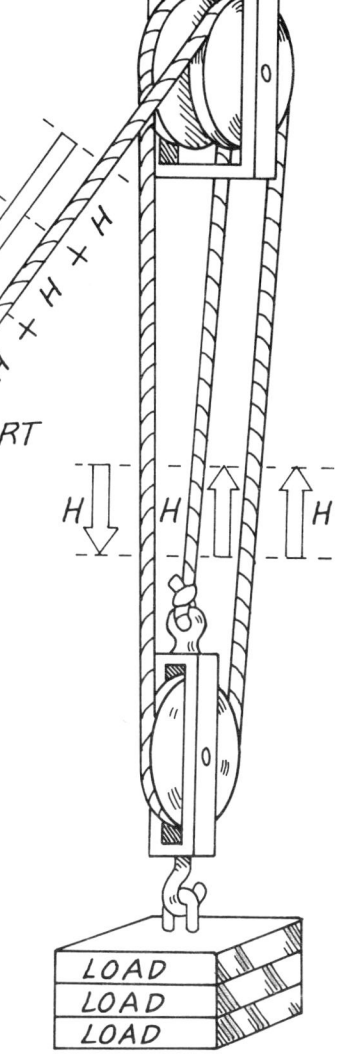

SAME EFFORT LIFTS THREE
TIMES THE LOAD

SPUR GEARS

When gear wheel **A** turns once, gear wheel **B** turns twice. This is because there are twice as many teeth on **A** as there are on **B**. This comparison of the number of teeth is called the gear **ratio** and in this case is 20 : 10 or simply 2 : 1. If **A** turns one hundred times in one minute, i.e. 100 **r.p.m. (r.p.m. are revolutions per minute)**, then **B** turns two hundred times in one minute, i.e. 200 r.p.m.

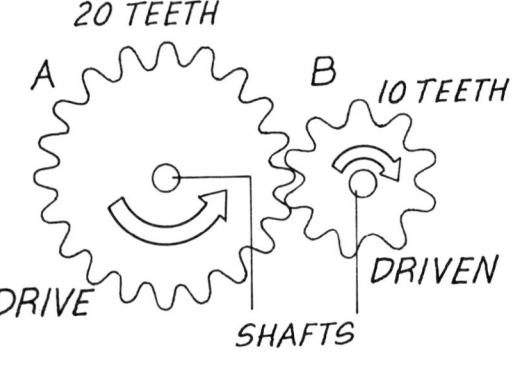

20 TEETH

A

B

10 TEETH

DRIVE

DRIVEN

SHAFTS

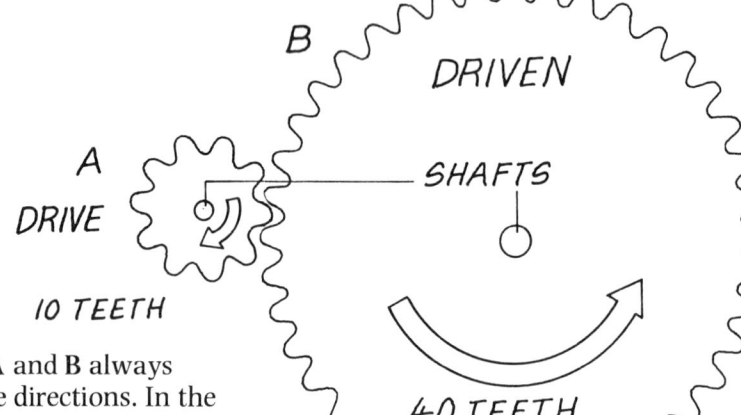

B

DRIVEN

A

SHAFTS

DRIVE

10 TEETH

40 TEETH

    Notice that **A** and **B** always turn in opposite directions. In the second drawing, you have **A** with ten teeth and **B** with 40 teeth. What is the gear ratio? If **A** turns at 100 r.p.m., at how many r.p.m. will **B** turn? When we step down the speed like this, shaft **B** can take a bigger load than shaft **A** (i.e. it takes more stopping). When we step up the speed it can only take less load. This is similar to mechanical advantage and velocity ratio (see Levers and Lifting Pulleys – Mechanisms 2 and 6).

GEAR TRAINS

Ratios
A:B = 1:4
C:D = 1:2

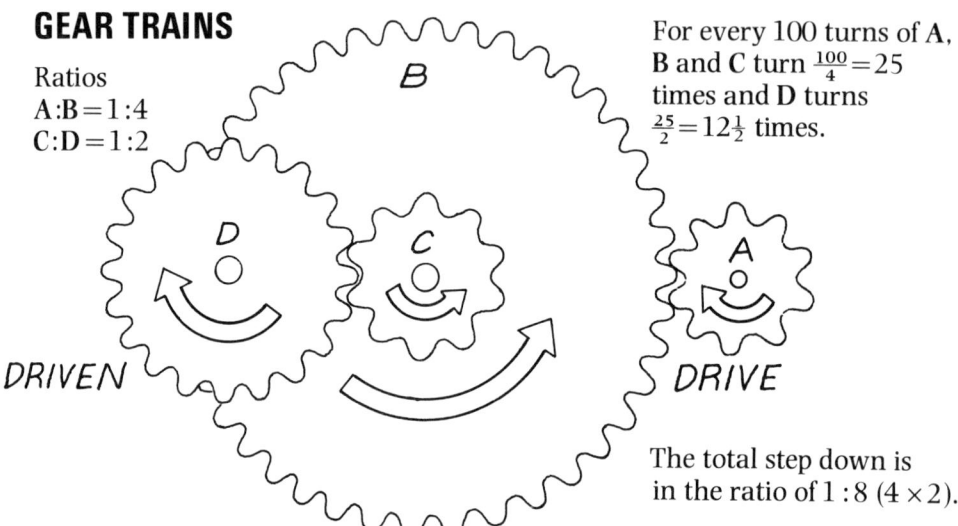

B

D

C

A

DRIVEN

DRIVE

For every 100 turns of **A**, **B** and **C** turn $\frac{100}{4} = 25$ times and **D** turns $\frac{25}{2} = 12\frac{1}{2}$ times.

The total step down is in the ratio of 1 : 8 (4 × 2).

When large changes of speed are required, several gears are needed. These are usually in pairs and one after the other (in series). This arrangement is called a gear train. (Note that **B** and **C** share, and are fixed to, the same shaft.)

IDLER GEARS

Ratio **A:B** is 1 : 1, i.e. the speed of **B** is the same as **A**.

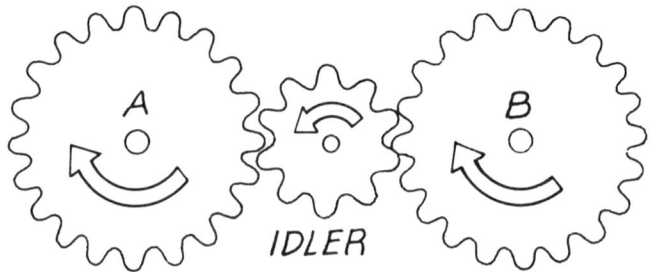

A

B

IDLER

    An idler gear is one used to make **A** and **B** both turn in the same direction. Note that the idler does *not* change the *speed* of rotation.

GEARS TO CHANGE THE MOVEMENT THROUGH 90°

Here are two Bevel Gears. These are gears where the teeth are angled at 45° to the edge of the wheel. Because the teeth can mesh at this angle the gear shafts can be at right-angles to each other.

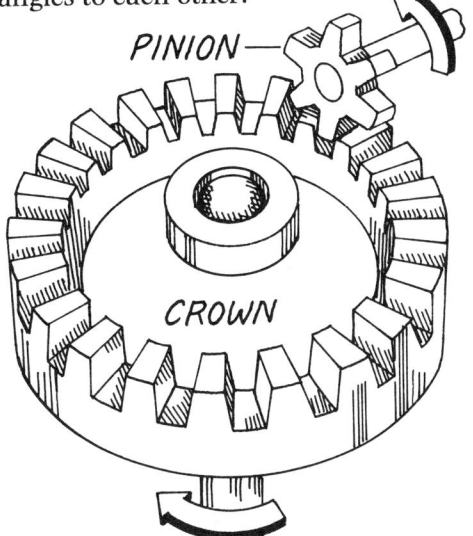

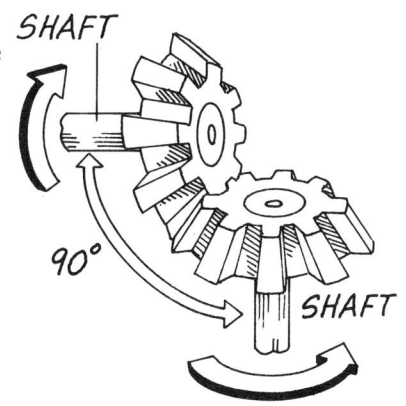

Here is a Crown and Pinion system. This is another way that you can change the angle of movement through 90°.

This system is called a Worm and Spur. The Worm is a gear with a spiral thread. The Spur must be driven by the Worm. It will not work the other way round. Because a Worm gear has to make one complete turn for every tooth on the Spur gear, it is possible to get very large reductions in speed just by using a Worm gear and a large Spur.

SOME OTHER USEFUL GEAR SYSTEMS

This arrangement is called a Rack and Pinion. It changes rotary movement into **linear** or linear to rotary. It is often used in steering mechanisms for cars and for opening the sluices on lock gates.

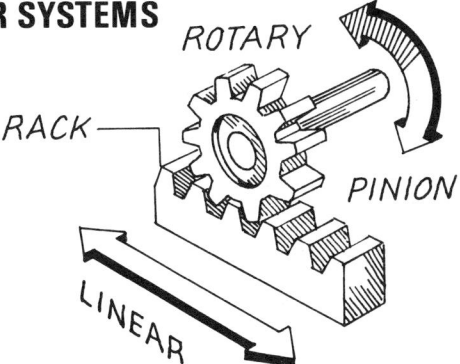

This is a Chain Drive. Here the teeth do not mesh into each other but are connected by a chain into which they mesh. Look at a bicycle. Unlike a Belt Drive, the linkage is very positive and there is no slipping.

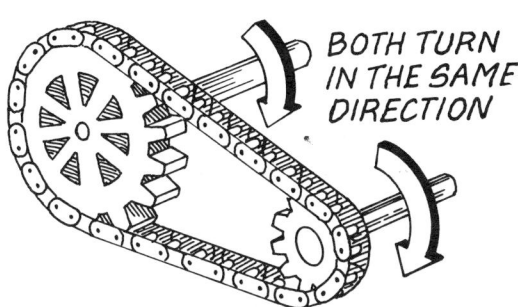

This device is called a Ratchet and Pawl. It will only allow rotation in one direction. This is because the Pawl can slide easily up one side of the Ratchet tooth but locks up against the other. **Windlasses** and winding mechanisms on clockwork often make use of this device.

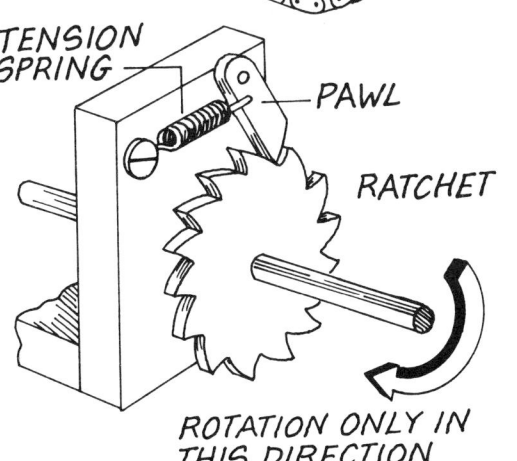

THE CRANK

Here is the 'starting handle' on old cars. This is the simplest form of the crank and is a **rotary** 'lever'.

This version of the crank converts rotary motion into **oscillating** motion using a slide. It does not change up and down oscillating motion into rotary. It is a one way system – perhaps you can work out why. This mechanism is useful for switching. (The slide can be used with the crankshaft shown on the left – how would this change the movement?)

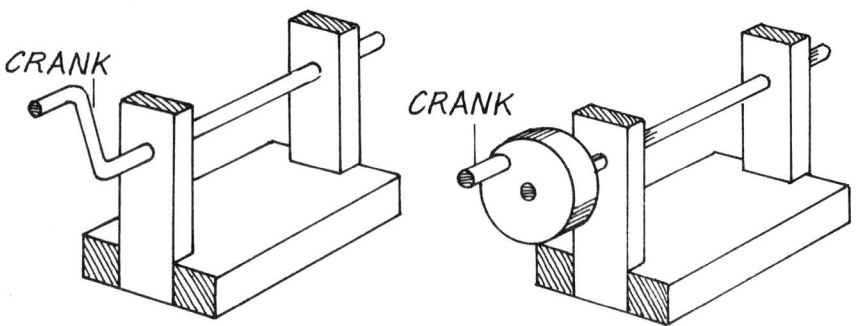

Here is a very useful form of the crank which either converts **rotary** motion into **reciprocating** motion (in and out) or reciprocating into rotary depending on which side of the system is powered. This **mechanism** is a vital part of the **internal combustion engine** (find out about cylinders, pistons and crankshafts).

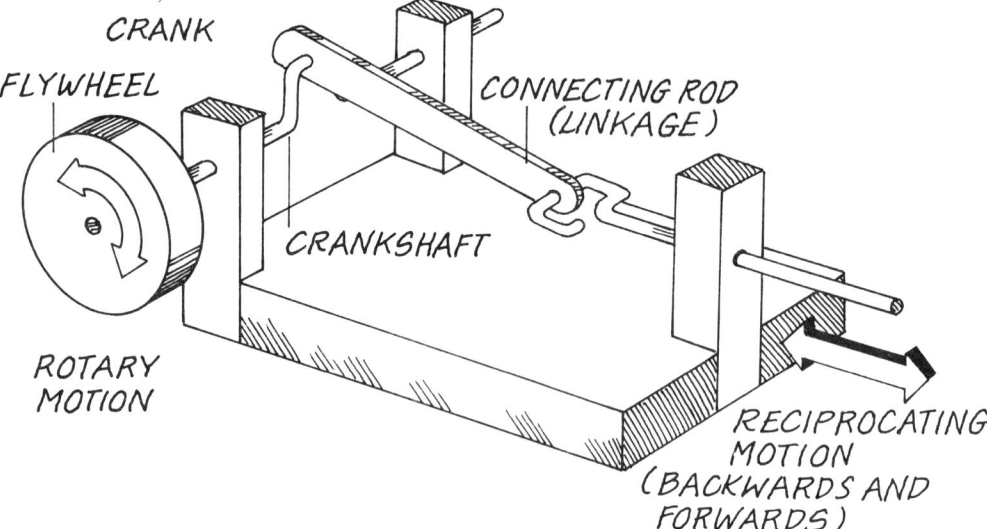

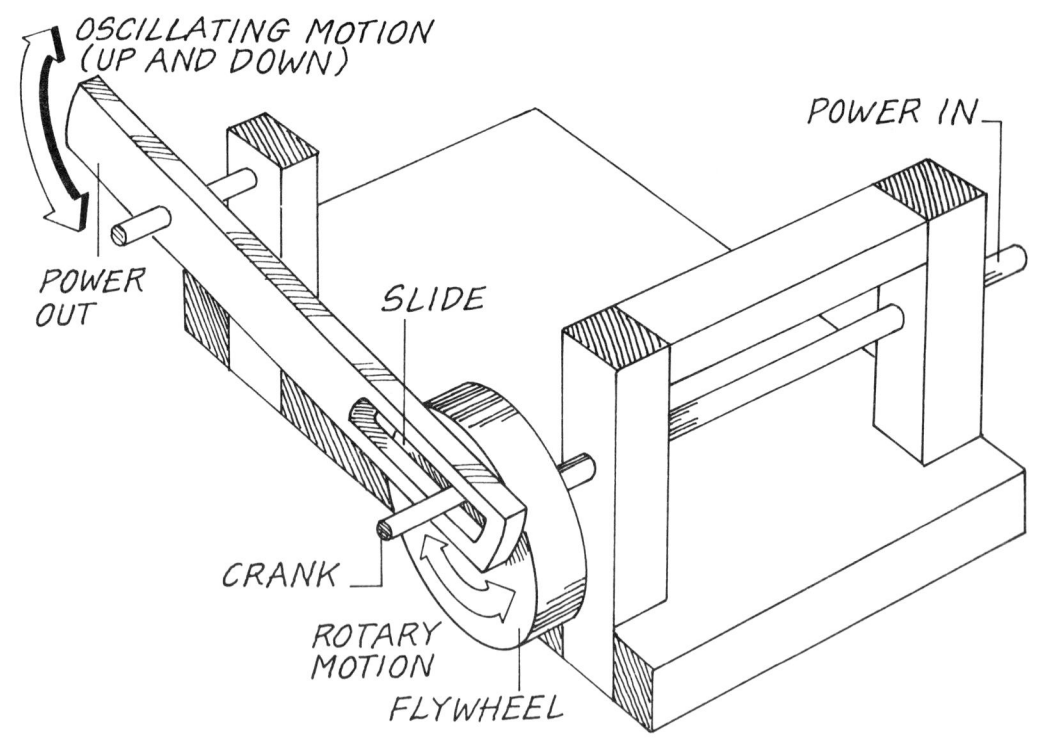

20

THE CAM

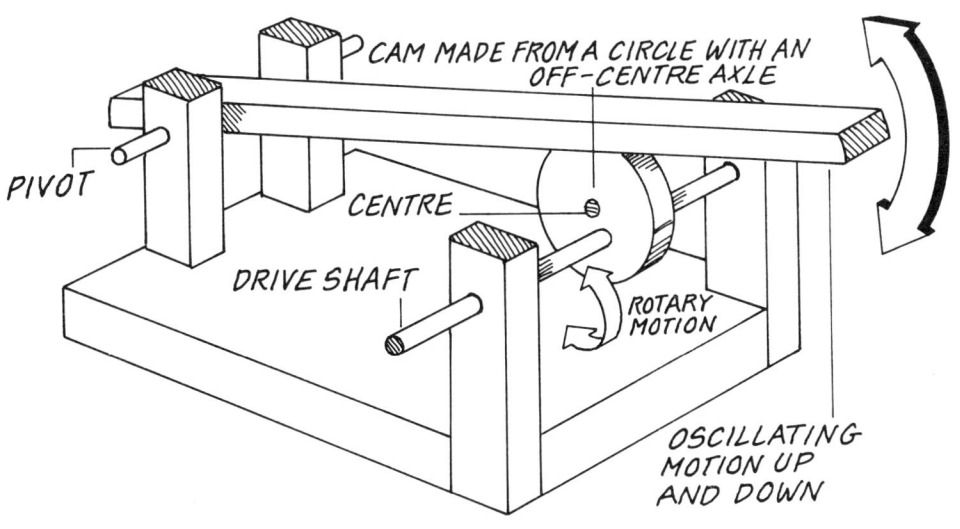

CAM MADE FROM A CIRCLE WITH AN OFF-CENTRE AXLE

PIVOT

CENTRE

DRIVE SHAFT

ROTARY MOTION

OSCILLATING MOTION UP AND DOWN

Here is a simple cam. This converts rotary motion into **oscillating** (up and down) but not the reverse. The cam shown is an off-centred circle and produces one up and one down movement for each complete **revolution**. By using different shaped cams many different kinds of movement can be produced.

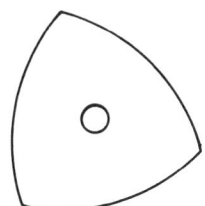

This produces three ups and three downs per revolution.

How many ups and downs will this shaped cam produce? What other shaped cams could be useful?

Cams can be used for counting the number of turns or part turns made.

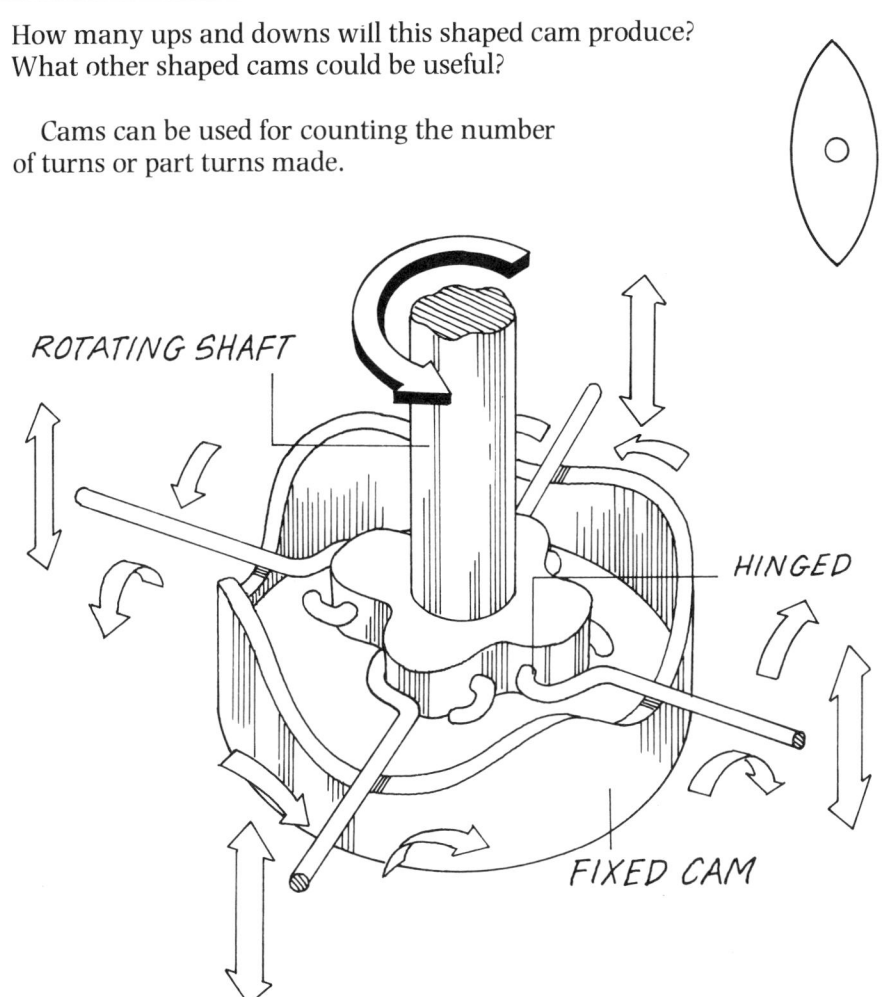

ROTATING SHAFT

HINGED

FIXED CAM

As the shaft rotates, the rods go round. They rise and fall, following the shape of the cam. If the shaft is fixed and the cam rotates how will the rods then move?

SINGLE AXLE STEERING

This is the simplest method of steering a vehicle (much favoured by children making carts!). It does, however, suffer from serious stability problems: i.e. it tips over very easily.

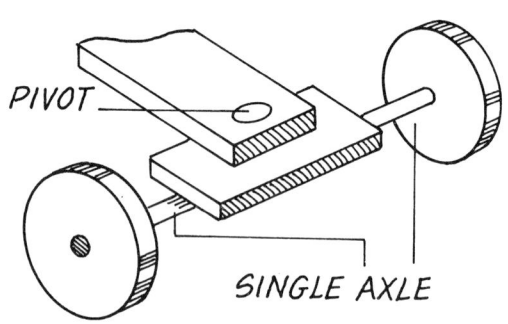

In the diagram below, the vehicle becomes unstable as the axle pivots and the front wheel on one side gets nearer to the back wheel. Another drawback is that the amount of movement the front wheels make in turning makes it difficult to design a body-shape that does not foul up by rubbing on the wheels. For these reasons this type of steering is only found usually on small trolleys and hand carts.

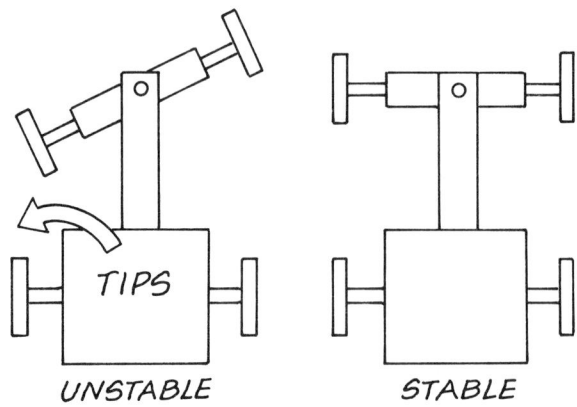

TWIN STUB AXLE STEERING

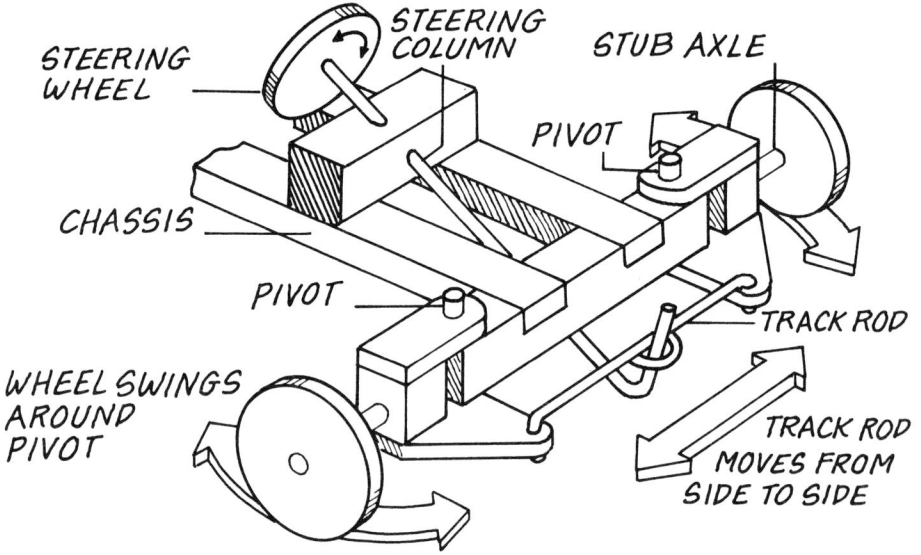

This kind of steering is a simple form of the one found in the motor car. Most of the problems of single axle steering are overcome because the wheels move through a much shorter distance. Turning the steering wheel causes the track rod to move from side to side which then causes each of the wheels and stub axles to turn about their pivots. Note that the wheels always stay parallel.

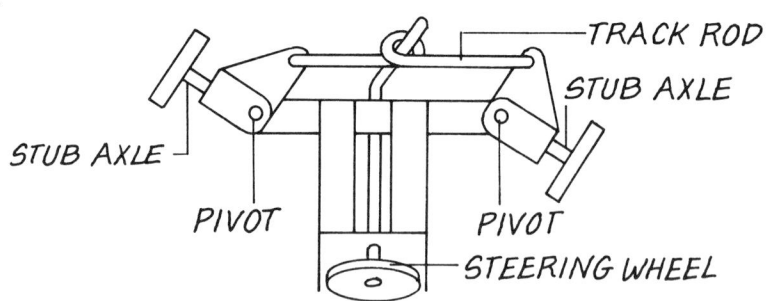

RACK AND PINION

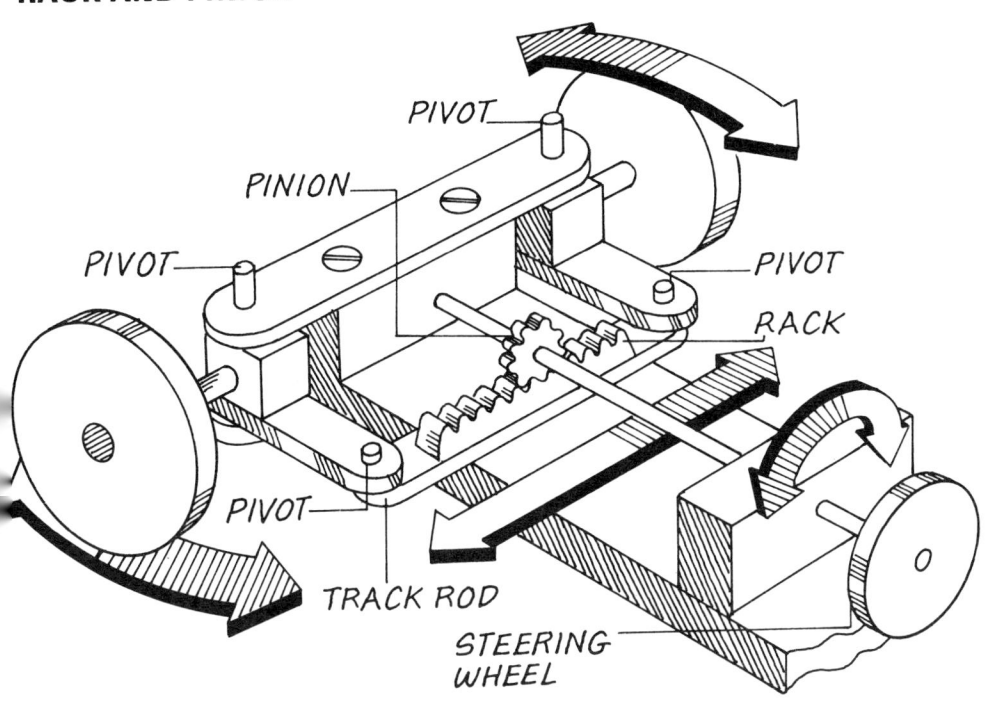

STEERING WITH TWO MOTORS

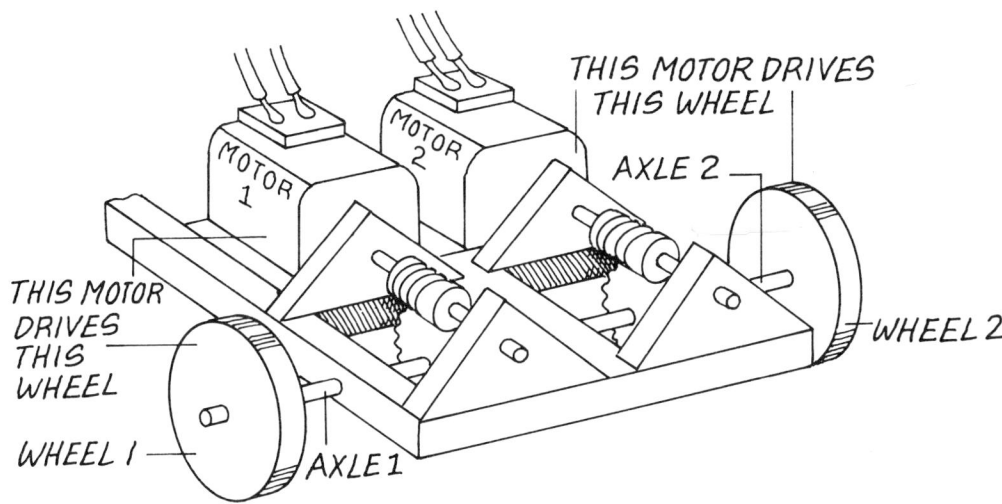

This method of steering is the one used in computer-controlled buggies. By switching both motors so the wheels both turn in the same direction the buggy goes either forwards or backwards. Switch motor 1 on forwards and motor 2 off; this time the buggy turns left. If motor 1 is switched off and motor 2 on forwards, the buggy turns right. By switching motor 1 on forwards and motor 2 on reverse, the buggy rotates left. By switching motor 1 on reverse and motor 2 on forwards, the buggy rotates right. Buggies do not usually have four wheels but have a ball caster instead of the rear axle.

Here is a simple rack and pinion steering mechanism. Turning the steering wheel turns the pinion and moves the rack. The rack is fixed to the track rod. The track rod pushes the pivots connected to a form of Bell Crank (Mechanisms 4) at each end. One arm of each Bell Crank forms the stub axle around which the road wheel **rotates**. Turning the steering wheel makes the road wheels swing parallel with each other around their pivots. We could replace the steering wheel with a motor, a gearbox (to reduce the speed) and a changeover switch (see Energy and Control 6). We then would have a simple remote control system. There would be several problems with this type of system. Can you see what they are?

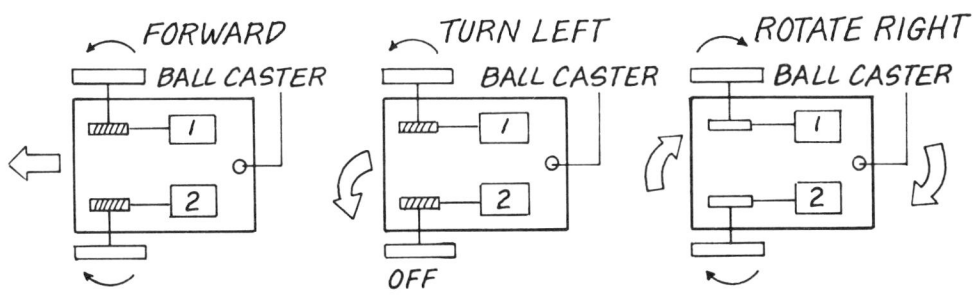

USING SCIENCE

Clockwise from left. Bus being tilted to test its stability; Formula 1 tyres; steel band pans being played; inshore lifeboat in action; skater.
Some of these pictures show what happens when some materials are rubbed together. What are the materials? Why are the results so different?

BALANCE   MOMENTS   EQUILIBRIUM

MOMENTS

Turning forces are called 'moments'. A moment depends on the size of the force and its distance from the turning point or pivot.

Moment = force × distance between force and the pivot

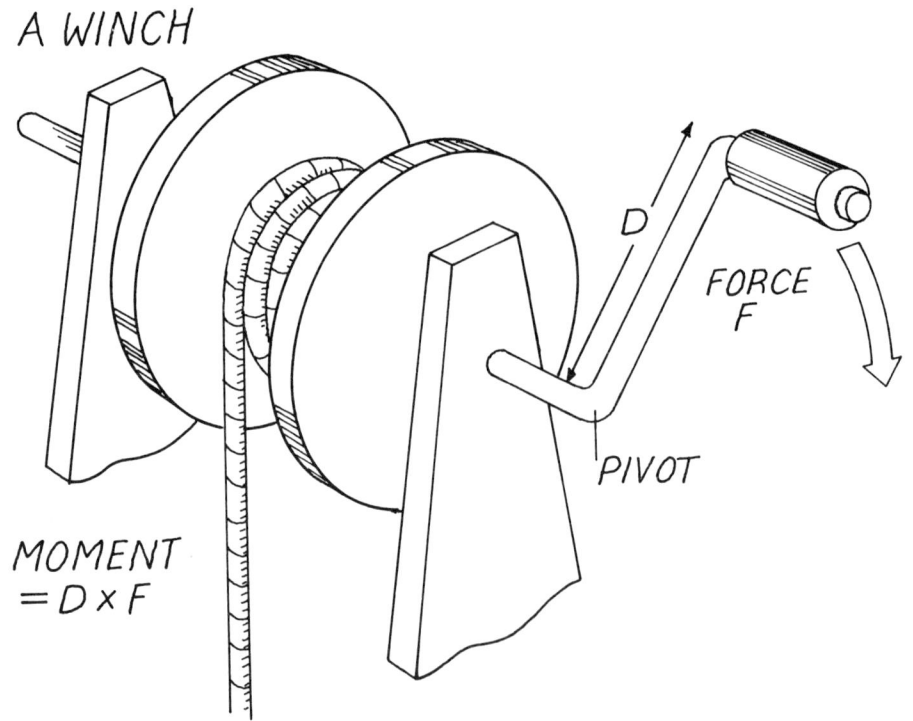

A WINCH

D

FORCE
F

PIVOT

MOMENT
= D × F

(**D** is the distance between force and pivot.)

EQUILIBRIUM

Imagine two people on a see-saw. If they are both exactly the same weight and exactly the same distance from the central pivot, the see-saw will balance. This condition of balance of opposite forces is called an equilibrium.

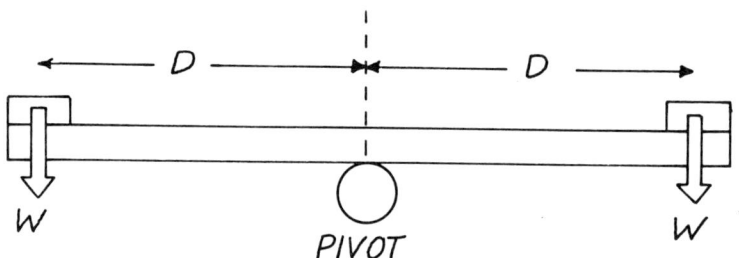

W   PIVOT   W

Left-hand moment = D × W = right-hand moment

If the left-hand person suddenly doubled in weight, the balance (equilibrium) would be destroyed. How could equilibrium be regained?

If the left-hand person (weight 2 × W) moves so he or she is now only half the distance (**D/2**) from the pivot of the right-hand person, then equilibrium will be restored.

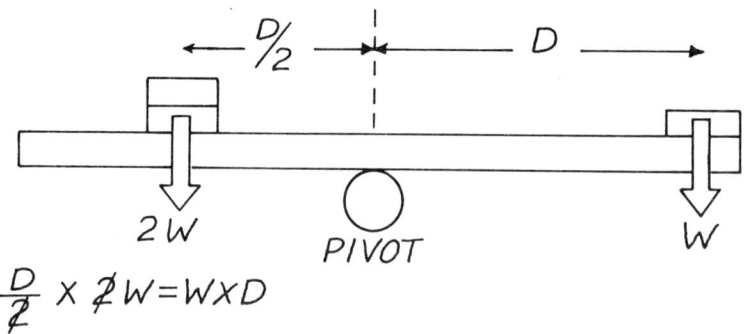

2W   PIVOT   W

$$\frac{D}{2} \times 2W = W \times D$$

For a beam in equilibrium, the clockwise moments are equal to the anti-clockwise moments.

This **principle** is often used in mechanical weighing machines.

WEIGHT

All objects have mass. Objects having mass are attracted towards each other. This attraction is known as the force of gravity. On Earth the force of gravity is dominated by the mass of the Earth. All objects on Earth are attracted towards the Earth. This attraction is what we call weight. The force of gravity pulls everything towards the centre of the Earth which is the Earth's centre of gravity. This is why the direction 'down' depends on where we are on the Earth's surface.

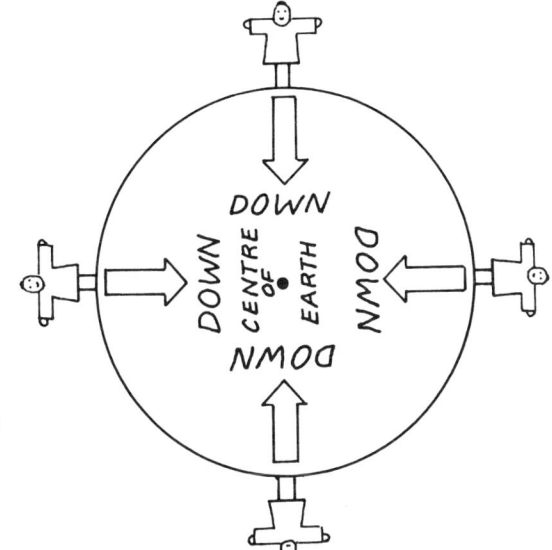

Which direction is down?

CENTRE OF GRAVITY

Just like the Earth, in which the force of gravity appears to be acting through its centre, every object has its own centre of gravity. The centre of gravity is taken to be the point from which the weight of an object appears to be acting. This idea is very important when we are looking at the stability or instability of a structure (i.e. is it likely to fall over or not?)

STABILITY

Which one of these two is more likely to fall over in a wind?

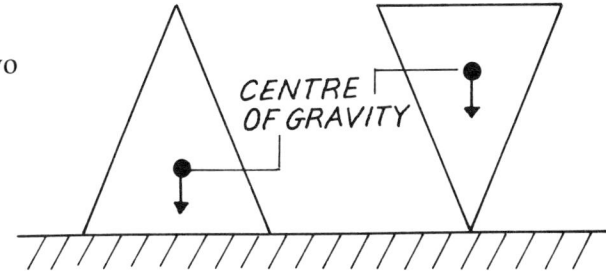

The one on the left is stable. The one on the right is unstable. This is because the one on the left has a wider base and the centre of gravity is low down. The weight acting through the centre of gravity, however, must also act through the base.

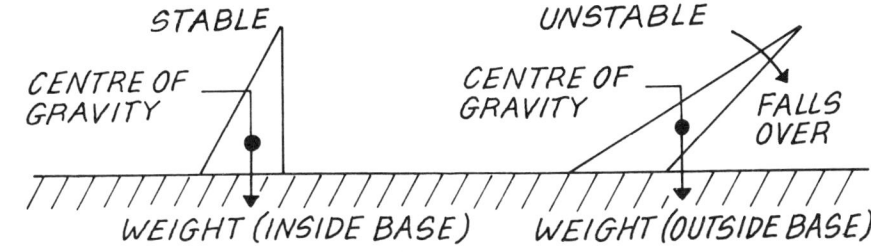

Most objects, however, will fall over if they are pushed far enough.

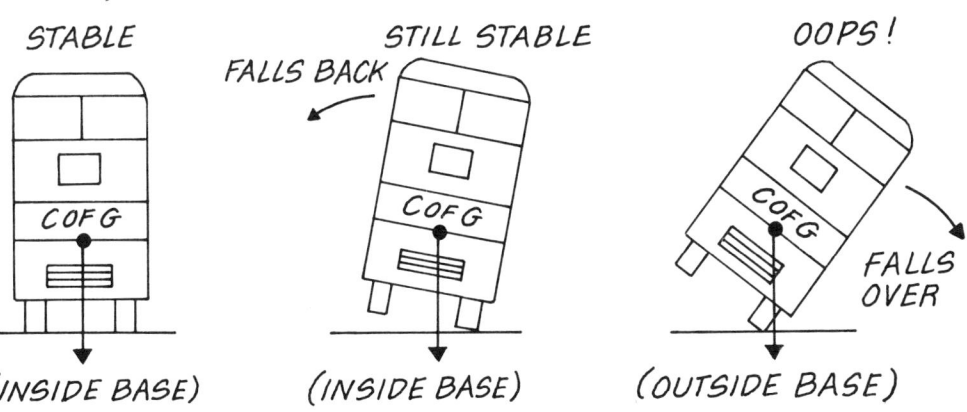

BUOYANCY   FLOATING   STABILITY ON WATER

BUOYANCY

Objects seem to weigh less when they are in a liquid than they do in air.

Look at the diagram on the right. Here the same object is weighed using a spring balance firstly in air then in water. The object in water appears to have lost weight. This apparent loss of weight is called buoyancy and is what allows objects to float.

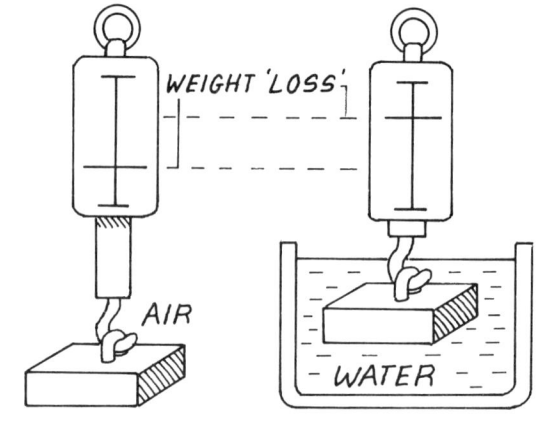

FLOATING

What happens to the level of the water in the beaker as the object is lowered in on the end of the spring balance? It rises. The volume of the water that has been lifted up is the same as the volume of the object. The weight of the water pushed out of the way (displaced) is equal to the apparent loss in weight of the object. This is called Archimedes' Principle. This '**displacement**' of water is very important when we want to know if an object will float.

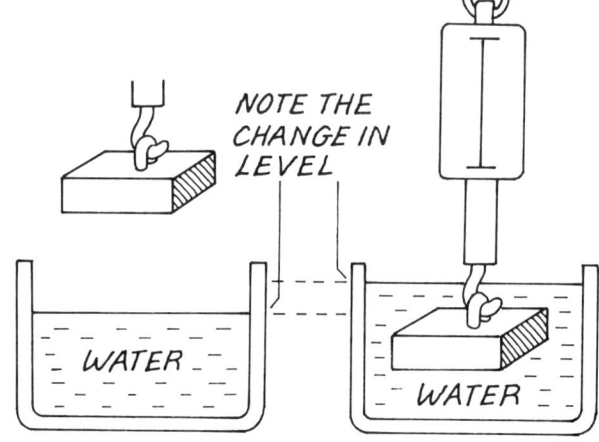

Here we have a container brim full with water resting in a deep tray. Put a margarine tub gently onto the surface of the water. A little water will overflow into the tray. Place a 10 g weight in the tub. More water overflows. Place further weights in the tub and notice what happens each time. Take the tub and the weights out of the water. What happens to the water level? Pour the water that has overflowed into the tray into a measuring cylinder. The number of millilitres will be the same as the number of grams of water (one millilitre of water weighs one gram). Compare this weight with that of the tub and weights. What do you find? If you compare the volume of the tub with the volume of the water displaced you will see that the volume of the tub is greater. Objects float when the weight of their total volume is less than the weight of the same volume of water.

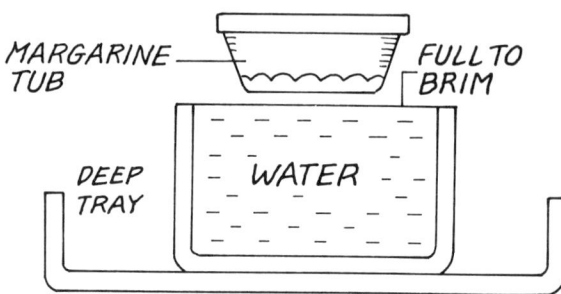

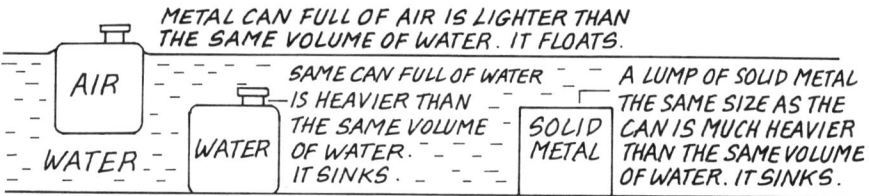

STABILITY ON WATER

Boats overturn (capsize) if their centre of gravity is too high above the water level. Weight is often added to boats in the form of keels or ballast to prevent this happening. Try to find out about the Plimsoll line.

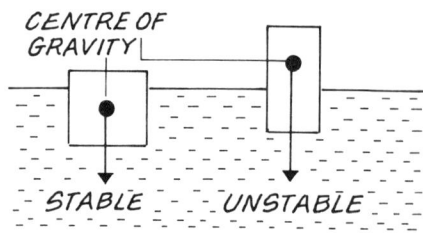

FRICTION

Friction is the resistance to motion which happens when we try to move one surface over another. Friction only exists when another force is acting. It is what we call a 'reactive' force. **Static** friction is the name given to the friction between two surfaces where a force is trying to slide one over the other but movement is not taking place. Kinetic (moving) friction is the name given to the friction caused when one surface does slide over the other. Rolling friction is the name given to friction between the surface of a rolling object and the surface on which it rolls.

  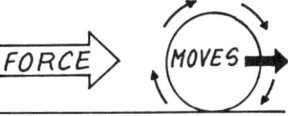

STATIC FRICTION   KINETIC FRICTION   ROLLING FRICTION

Usually, for any two surfaces, static friction is greater than kinetic friction. This means that, once an object starts sliding, the force needed to keep it sliding is less than the force that first caused it to slide! Rolling friction is less than sliding (kinetic) friction. That is why vehicles in countries without snow and ice have wheels rather than runners. (Why might snow and ice make runners useful?) Friction, though a nuisance in some cases, is vitally important in others. For instance, although we want friction to be very low on a wheel bearing, we want friction to be very high where the wheel touches the ground. Without friction here we would have very poor **traction** and so would not get very far. It would be like trying to ride a bike with bald, wet tyres on smooth, wet ice! There are times when we would like to increase friction as well as times we would like to reduce it. As well as tyres, think of brakes, ladders against walls and the soles of our shoes.

WE WANT LOW FRICTION IN THESE BEARINGS

WE WANT HIGH FRICTION IN THESE PLACES

INCREASING FRICTION

The rougher a surface is, the greater the friction. So one way to increase friction is to roughen the surfaces in contact. In the diagram the block covered in sandpaper does not slide as easily as the block without. Rubber will work in the same way, as will any substance that makes the surface sticky or increases the resistance to sliding.

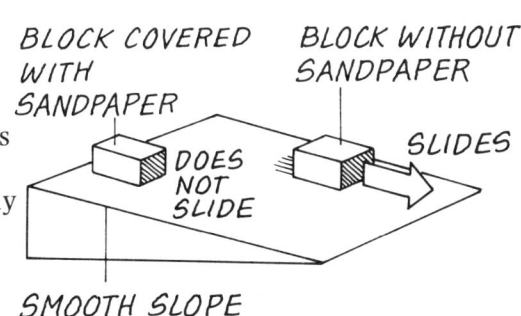

BLOCK COVERED WITH SANDPAPER   BLOCK WITHOUT SANDPAPER

DOES NOT SLIDE   SLIDES

SMOOTH SLOPE

REDUCING FRICTION

Choose materials which have slippery surfaces, i.e. polished metal on polished metal has much less friction than wood on wood (or wood on metal). Use substances which help the surfaces to slip over each other. These are called lubricants, e.g. oil, wax, graphite (pencil lead). What about ball bearings?

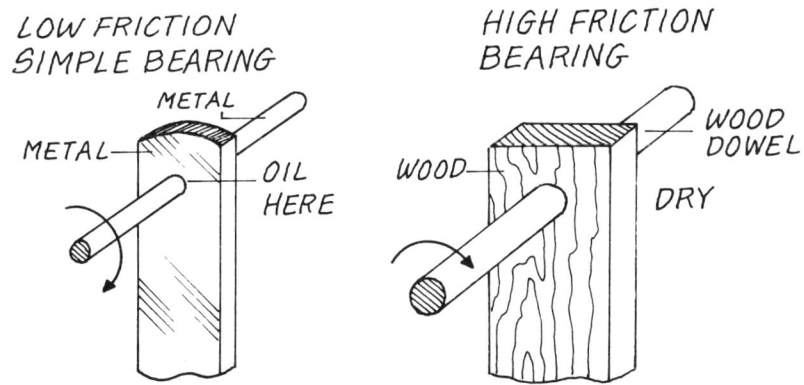

LOW FRICTION SIMPLE BEARING   HIGH FRICTION BEARING

METAL   METAL   OIL HERE

WOOD   WOOD DOWEL   DRY

ELASTICITY   HOOKE'S LAW   SPRING BALANCES   USING BENDING – TORSION

ELASTICITY

All solid materials show a certain flexibility and will bend or stretch when a large enough force is used on them. If the material goes back to its original shape when the force is removed, then the material is said to be elastic. There is a limit to how far a material can be stretched and still return to its original shape. This is called the elastic limit.

Materials which do not return to their former shape when the force is removed are said to be inelastic.

A rubber band is elastic.

HOOKE'S LAW

This says that as long as the material is not stretched beyond its elastic limit, the change in length is **proportional** to the force which has made that change. This means doubling the force will double the change in length until the elastic limit is reached.

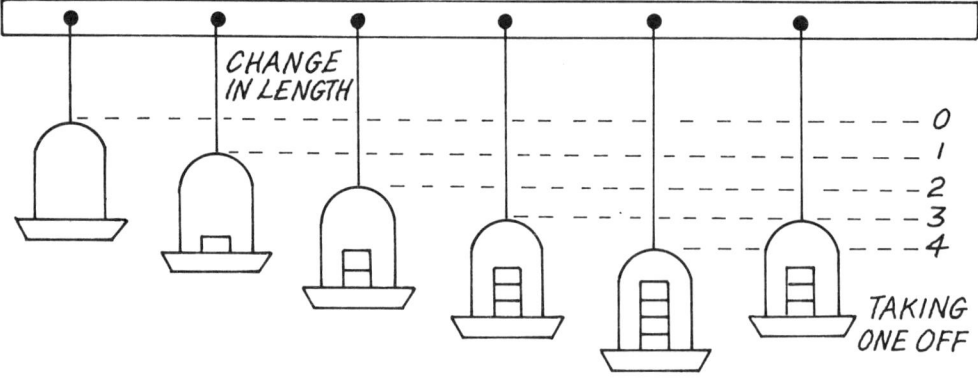

Each added weight increases the length by the same amount.

SPRING BALANCES

On the right is a very simple (and not very accurate!) rubber band balance. By choosing your rubber band with care you should get the sort of change in length you need for the sizes of weights you are using. How does this compare with a real spring balance? Try using a small spring instead of the rubber band. Is it more accurate? Which would you choose?

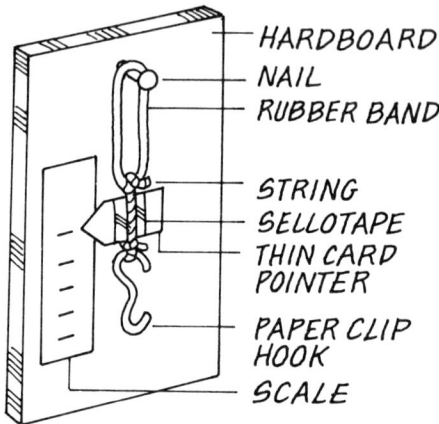

USING BENDING

Here is another way of using elasticity for weighing. When we add weights to the hook the hacksaw blade bends. The larger the weight the more it bends.

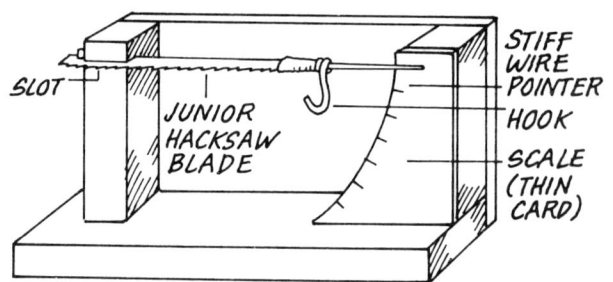

USING TORSION

Torsion means twisting. Adding weights to the hook increases the twisting of the rubber tube. The bigger the weight the greater the twist shown by the movement of the pointer.

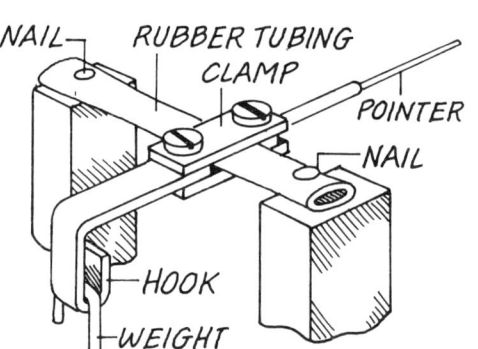

SOUND

Sound is the name we give to **vibrations** in the air (or in other substances). Anything which vibrates will cause the surrounding air to vibrate. These vibrations in the air spread outwards rather like ripples in a pool into which something has been dropped. Sound does not travel in a **vacuum**. It travels more quickly and further in solids than it does in air. Sound travels through air at about 340 m per second.

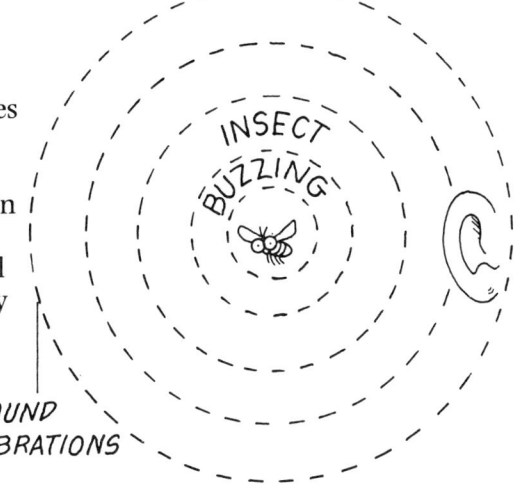

INSECT BUZZING

SOUND VIBRATIONS

MAKING SOUNDS

The simplest way of making a sound is by tapping one solid object against another. On the right we have hanging metal tubes of different lengths. These are called tubular bells. The shorter the tube the higher the note.

Other examples of sounds made by tapping are the triangle, the slit drum and the xylophone.

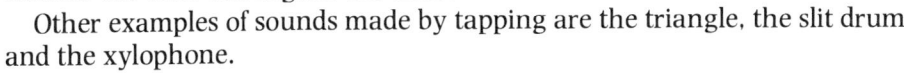

HANGING METAL TUBES OF DIFFERENT LENGTH

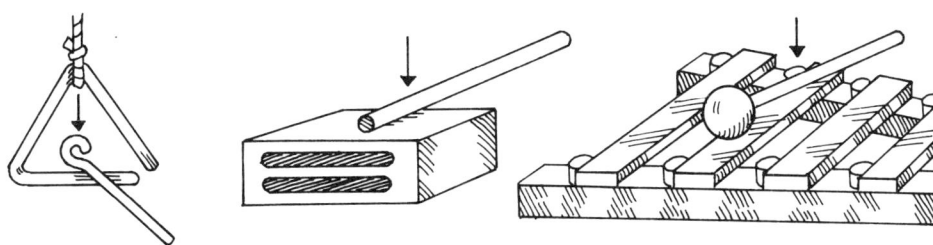

Stretch thick plastic sheets tightly over the mouth of a large tin and tie string round the sheet to keep it in position. This will make a simple drum. Banging this will make a loud noise.

Another way of making a sound is by using a rattle. If you put a few dried peas inside an empty washing-up liquid bottle or other suitable container, and shake it, you will have a simple rattle.

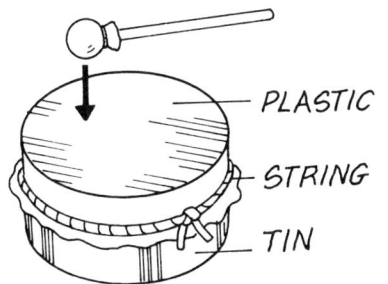

PLASTIC

STRING

TIN

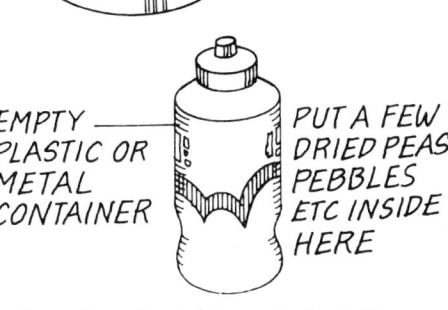

EMPTY PLASTIC OR METAL CONTAINER

PUT A FEW DRIED PEAS. PEBBLES ETC INSIDE HERE

Cramp a ruler or a hacksaw blade onto the edge of a table and pluck it. This too can produce quite a loud noise. You can make this noise louder by fixing the vibrating blade (or blades) onto a large hollow box.

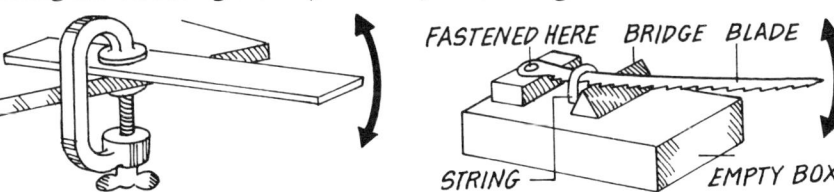

FASTENED HERE   BRIDGE   BLADE

STRING

EMPTY BOX

This is a football rattle. Can you see how the cog wheels make the thin strips of wood vibrate as the head of the rattle swings round and round?

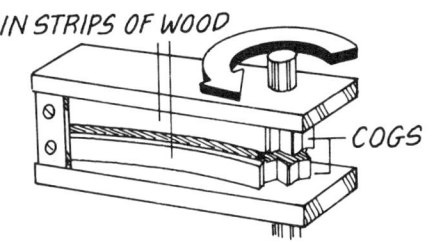

THIN STRIPS OF WOOD

COGS

ENERGY AND CONTROL

Clockwise from left. Electromagnet used for carrying metal scrap; sailing, Americas Cup 1986; cuckoo clock with falling weights; model aeroplane with an elastic-powered airscrew/propeller; wind turbine generator. **What is the source of energy in each case?**

GRAVITY POWER   FALLING WEIGHT   FALLING FLUID

FALLING WEIGHT

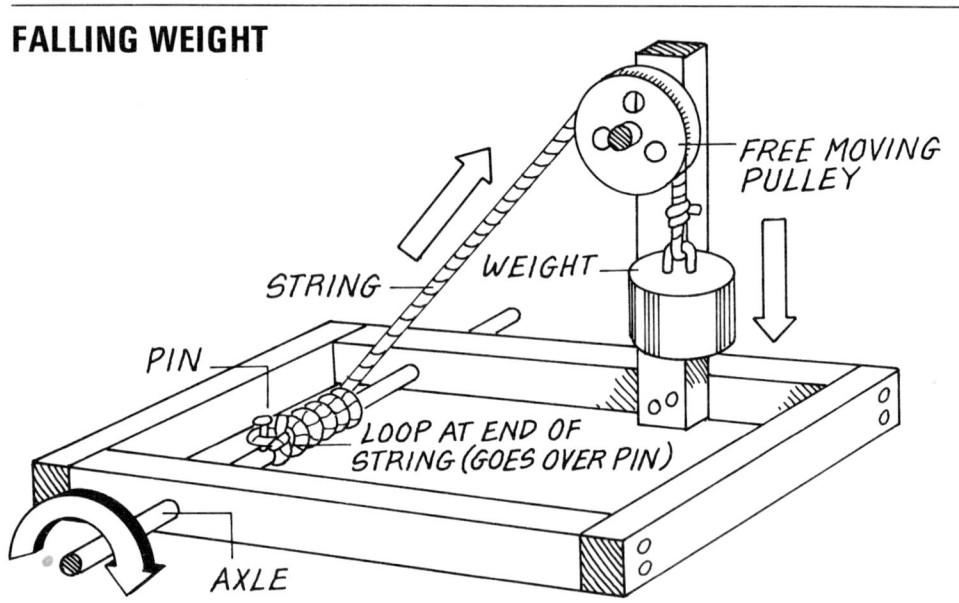

As the weight falls it pulls the string over the **pulley**. The string unwinds from round the **axle** making the axle turn as it does so. When the string reaches the loop at the end, this slips off the pin leaving the axle free to continue turning. This is called 'freewheeling'.

The arrangement on the right uses exactly the same **principle** but produces motion about the **vertical axis**.

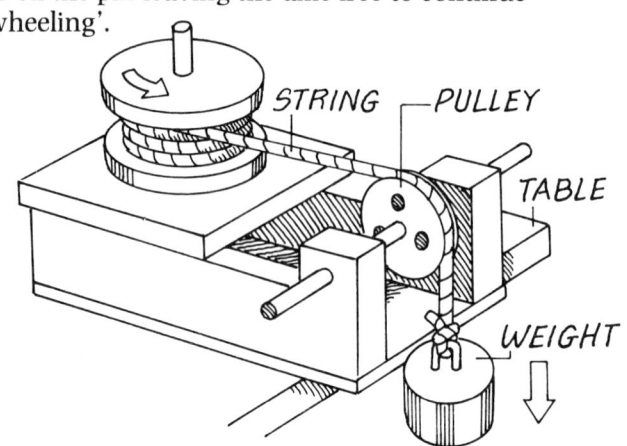

FALLING FLUID

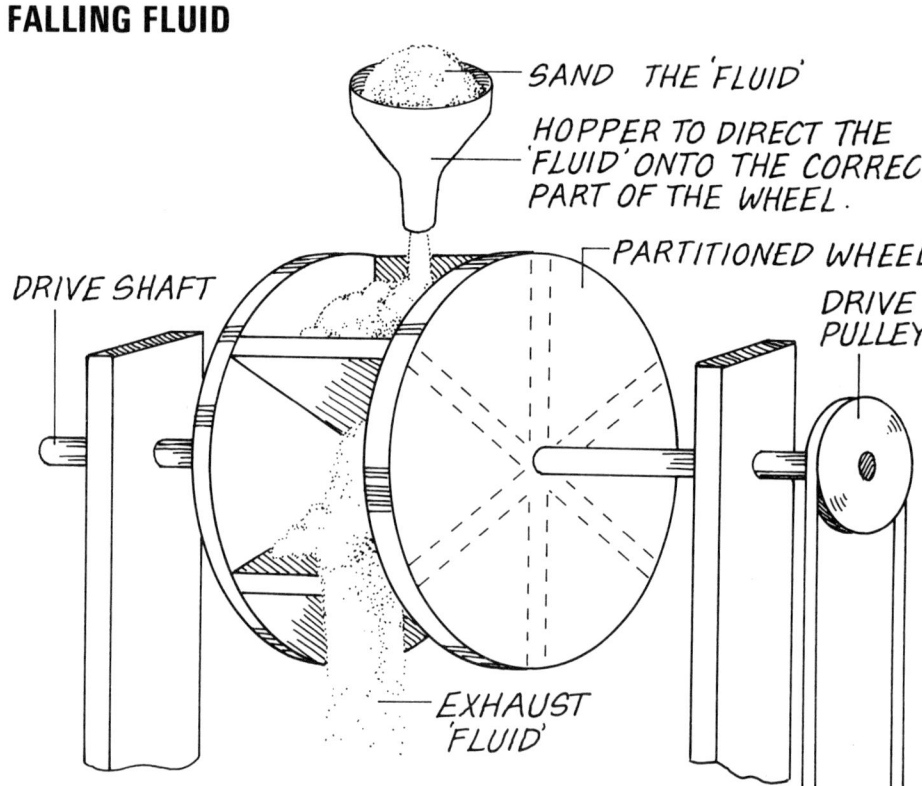

The **hopper** directs the dry sand (fluid) onto one side of the wheel so filling up one of the **partitions**. This changes the balance of the wheel because this side is now heavier than the other. This heavy partition falls sideways and downwards making the wheel and axle turn. This brings another partition under the hopper. What happens to the sand as the wheel tips further?

The fluid acts almost as fuel. What happens when the hopper is empty?

What do we call this **device** when water is the fluid?

Perhaps the easiest way of using gravity is rolling downhill! Can you think of examples?

WIND POWER

SAILS

The easiest way to make use of the wind is by using a sail. Although it is easy to move in the direction of the wind (to sail downwind) we don't always want to go that way. To sail at 90° to the wind is more difficult and sailing towards the wind is very difficult indeed. What other disadvantages to wind power can you think of?

SQUARE SAIL

The square sail is best for sailing in or near to the same direction as the wind.

MAST
SPAR
SAIL
TAIL WIND
SAILING DIRECTION

LATEEN SAIL

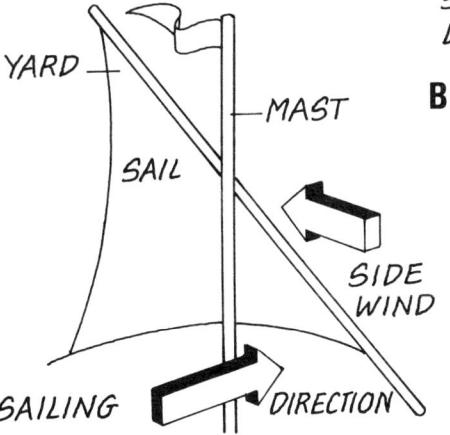

YARD
MAST
SAIL
SIDE WIND
SAILING DIRECTION

The lateen and Bermudan sails are much better for sailing into a wind than the square sail. With these sails a boat (or vehicle) will need a low centre of gravity. This means usually a deep heavy keel in the case of a boat.

BERMUDAN SAIL

MAINSAIL
MAST NEAR HEAD
WIND
FORESAIL LAPS BEHIND MAINSAIL
BOOM
HINGE
SAILING DIRECTION

ROTARY POWER

Wind being a **linear** force and machinery being mainly **rotary**, it would be convenient if we could change the wind power directly into rotation.

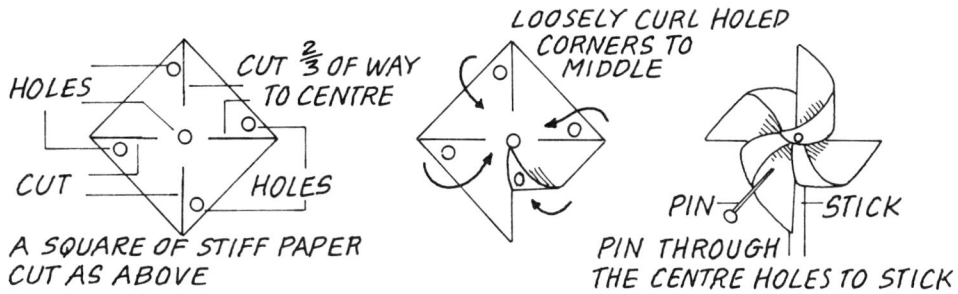

HOLES
CUT ⅔ OF WAY TO CENTRE
LOOSELY CURL HOLED CORNERS TO MIDDLE
CUT
HOLES
PIN
STICK
PIN THROUGH THE CENTRE HOLES TO STICK
A SQUARE OF STIFF PAPER CUT AS ABOVE

Here is a simple windmill often used as a child's toy which does the job we require. How could we make use of it? Here are two more ideas.

A SAIL ROTOR   ### SAVONIUS ROTOR

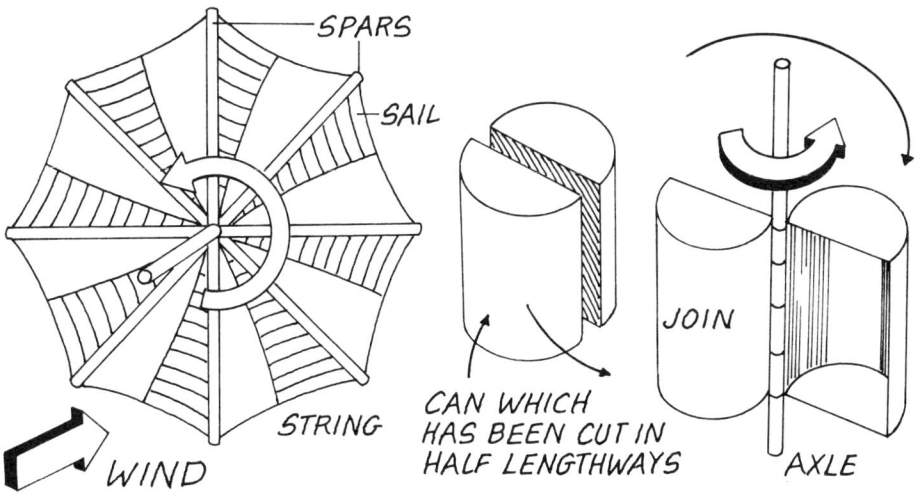

SPARS
SAIL
STRING
WIND

CAN WHICH HAS BEEN CUT IN HALF LENGTHWAYS
JOIN
AXLE

RUBBER POWER

STRETCHED ELASTIC

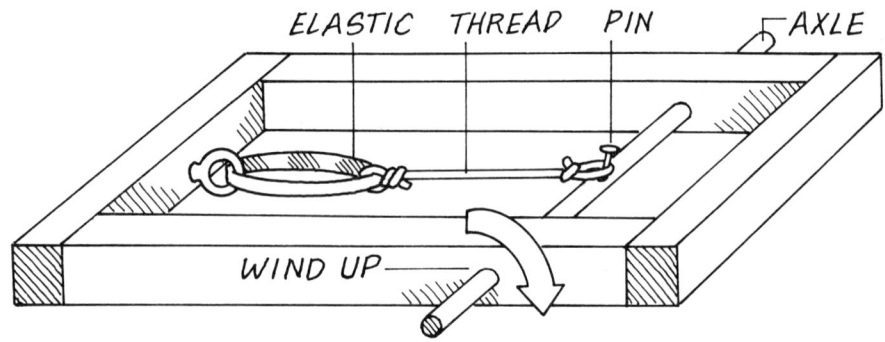

If the **axle** on the above frame is turned round and round in the direction shown the thread will be wound onto the axle and the elastic band will become stretched longer and longer, as shown below.

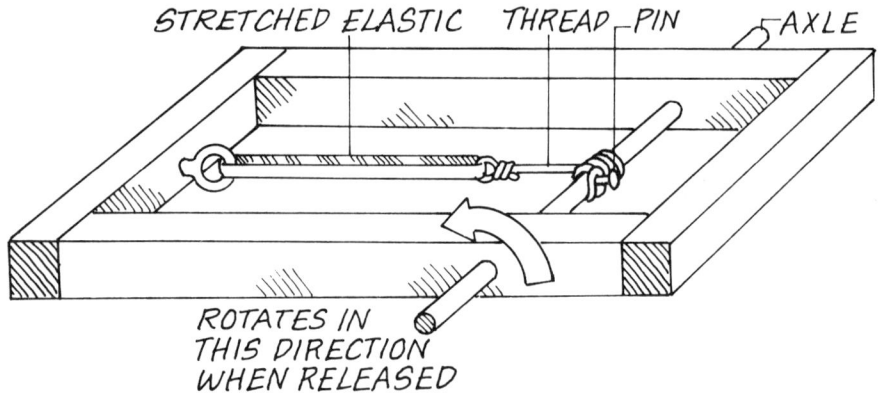

When the axle is released it will rotate rapidly in the opposite direction to which it was wound. If the thread is the right length, it will fly off the pin at the last moment. This will allow the axle to continue spinning under its own **momentum**.

    Where does the energy come from?

TWISTED ELASTIC

Cotton reel motors

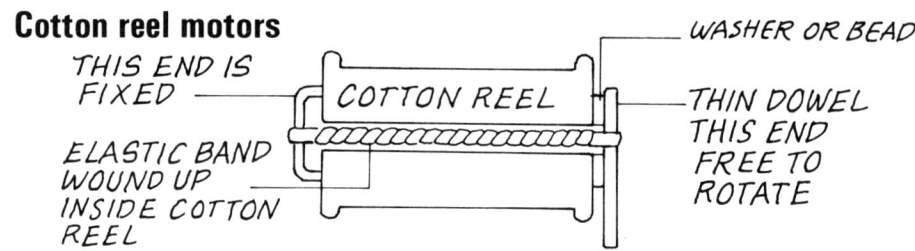

Wind the dowel end up until the elastic band is tight. Put the cotton reel on a flat surface and release the dowel. The cotton reel should now shoot off at high speed. The same principle could be used using a ringcan for the cotton reel and a longer elastic band and dowel.

USING AN AIRSCREW

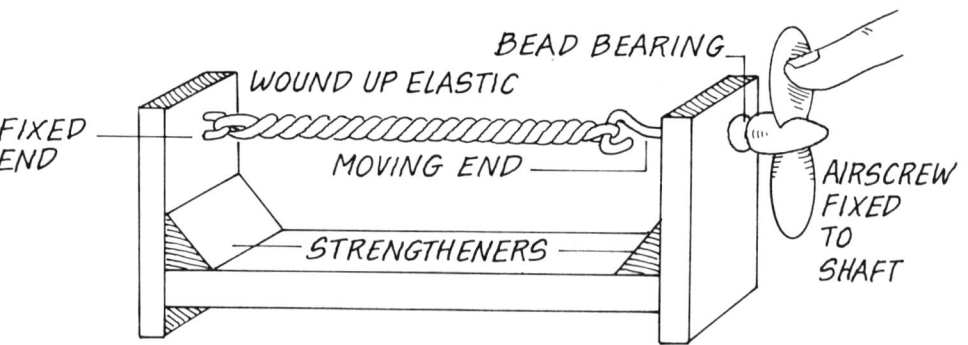

Remove the finger from the **airscrew** and it will rotate rapidly driving itself through the air (if the whole **assembly** is lightweight enough and free to move). Note that it is exactly the same as the cotton reel version with the propeller replacing the dowel.

REMEMBER: the elastic band does not make energy – it only stores up what *you* put into it in the first place.

BATTERY POWER

SINGLE CELLS

These are made in various sorts, shapes and sizes. Here is a typical dry cell. Although of different physical sizes, most single cell batteries have the same electron moving force (e.m.f.) measured in volts. This voltage is usually 1.5 volts (V) or less. Like all batteries, you need to make two connections, one to the **positive** (+) side and one to the **negative** (−) side, before anything can happen.

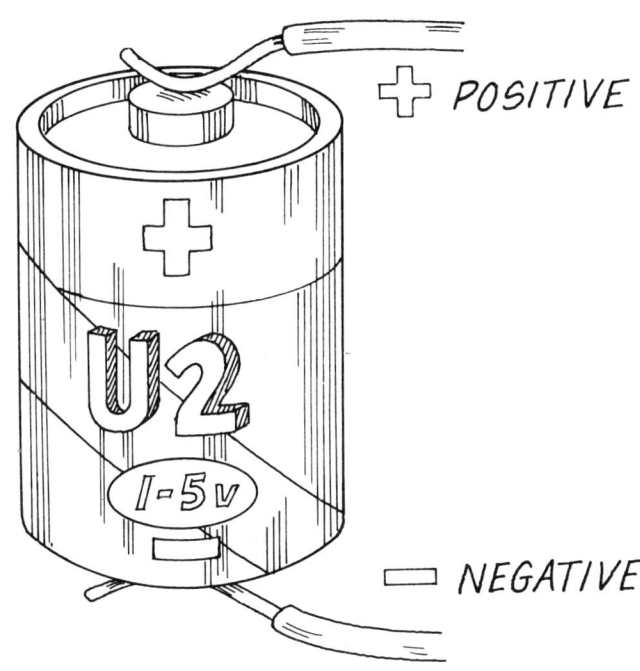

➕ POSITIVE

U2

⊟ NEGATIVE

When more than one cell is used, this is called a battery. The cells are usually connected together so they will produce a greater voltage. The method of connection is called 'series' and means one after the other.

BATTERIES

How to connect up single cells to make a 4.5 V battery

More cells can be added to the battery – each extra cell will increase the total voltage by 1.5 V.

How many 1.5 V cells are needed to make a 9 V battery?

Here is a 4.5 V dry battery. How many single 1.5 V cells does it contain?

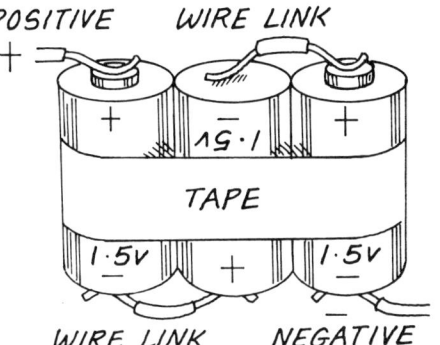

POSITIVE   WIRE LINK

TAPE

WIRE LINK   NEGATIVE

USING BATTERY POWER

Lighting a 1.5 V bulb

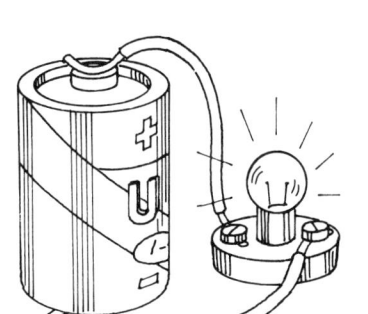

Driving a 4.5 V motor

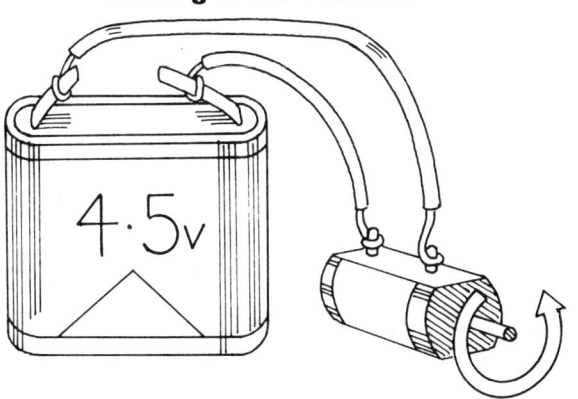

MOTORS AND BATTERIES

To use a motor with a battery it must be the correct type. This means it should:

1 be a **direct current** (d.c.) motor:
2 have a maximum voltage (rating) which is greater than or equal to the battery voltage.

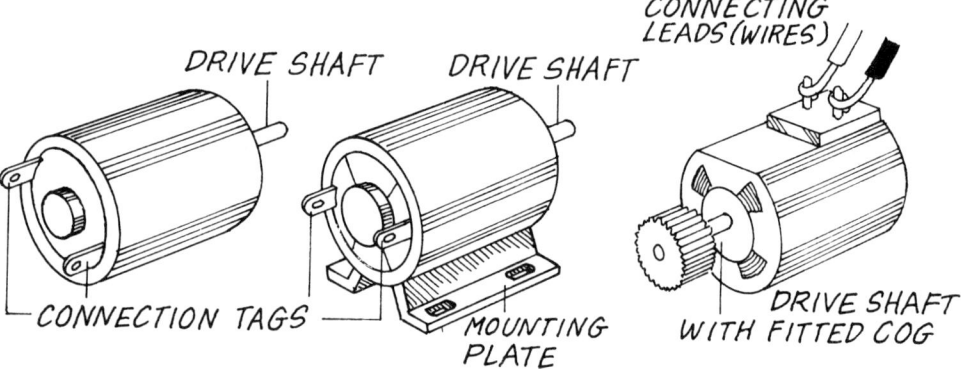

DRIVE SHAFT

DRIVE SHAFT

CONNECTING LEADS (WIRES)

CONNECTION TAGS

MOUNTING PLATE

DRIVE SHAFT WITH FITTED COG

Above are some typical d.c. motors.

The motor will have two connection tags or wires (leads). Connect a lead from one of the motor connections to the positive side of the battery (it must be the *metal* part of the wire not the plastic!) and the other connection to the negative side. The drive shaft of your motor will turn clockwise or anticlockwise.

Note which way the drive shaft is turning.

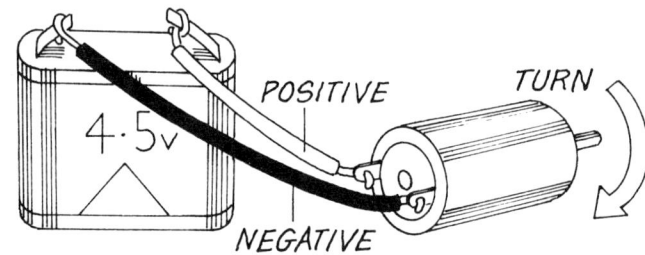

POSITIVE

TURN

4.5v

NEGATIVE

Now reverse the connections to the battery.
Which way does it turn?
How can you make use of this fact?
(See Motors and Switches – Energy and Control 6).

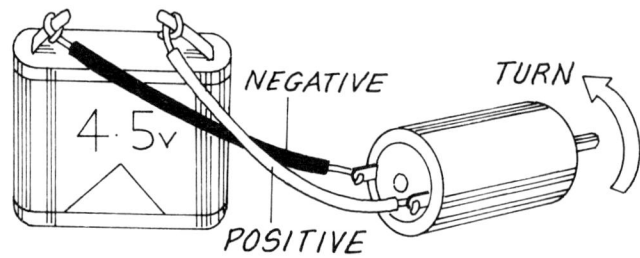

NEGATIVE

TURN

4.5v

POSITIVE

USING A MOTOR

Motors need to be mounted rigidly so that their drive shafts stay in the correct position and are not pulled or pushed out of **alignment**. The alignment, at right-angles or **parallel**, will depend on the method chosen to transfer the power from the motor drive shaft to the driven shaft (or lay shaft). This is called the **transmission**. (See Mechanisms 5, 6, 7 and 8.)
Some examples:

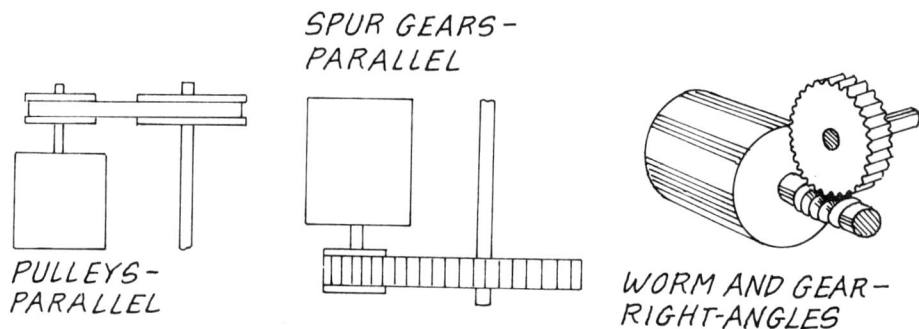

SPUR GEARS – PARALLEL

PULLEYS – PARALLEL

WORM AND GEAR – RIGHT-ANGLES

MOTORS AND SWITCHES

ON/OFF

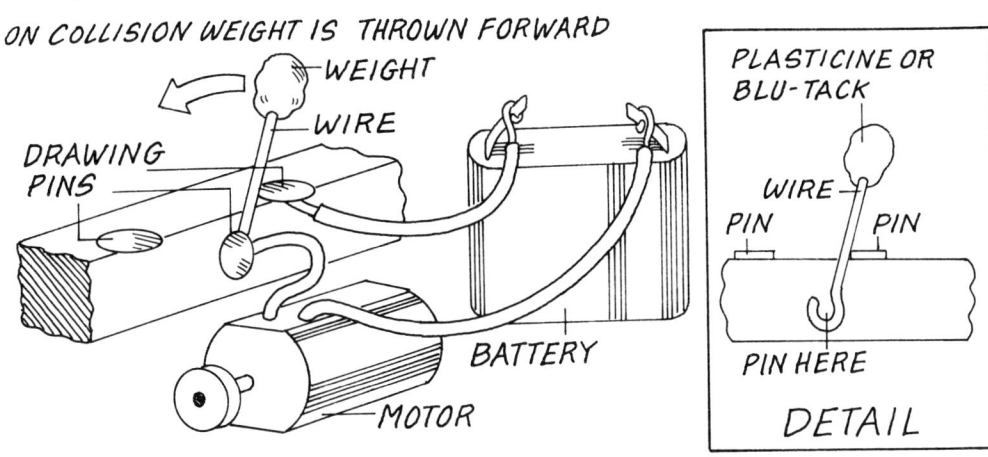

ON COLLISION WEIGHT IS THROWN FORWARD

WEIGHT
WIRE
DRAWING PINS
BATTERY
MOTOR

PLASTICINE OR BLU-TACK
WIRE
PIN — PIN
PIN HERE
DETAIL

By carefully setting the drawing pins and arranging 'on' to be with the weight leaning backwards, the above system can be made to switch off the motor if your vehicle collides head-on with an object. This is called an **inertia** switch.

If you mount a small 'popper' switch on the front of your vehicle it will switch off the motor if your vehicle collides with anything. You could mount the switch behind a hinged bumper.

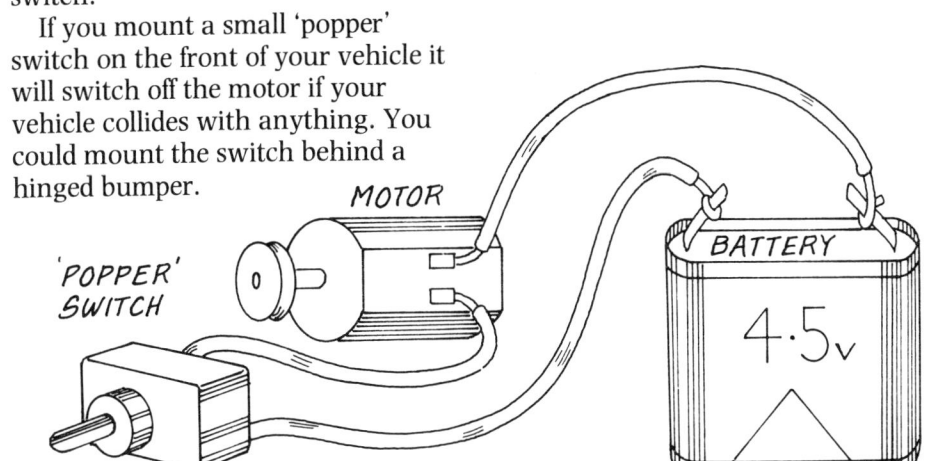

'POPPER' SWITCH
MOTOR
BATTERY
4.5v

FORWARD-OFF-REVERSE

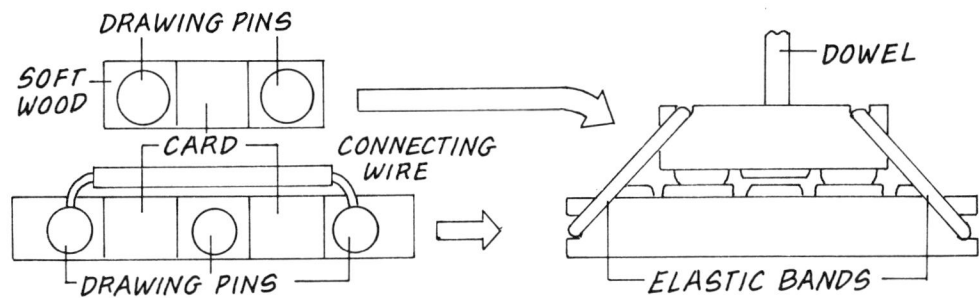

DRAWING PINS
SOFT WOOD
CARD
CONNECTING WIRE
DRAWING PINS
DOWEL
ELASTIC BANDS

You can make a simple forward–off–reverse (f–o–r) switch for yourself as shown above and wire it up as shown below right.

Make sure that the wire trapped under the drawing pins has had the **insulation** (the plastic coating) removed, otherwise it will not work.

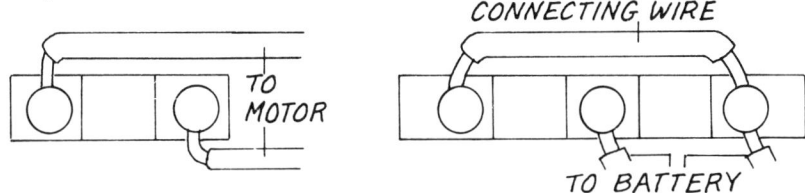

TO MOTOR
CONNECTING WIRE
TO BATTERY

WIRING UP A CHANGEOVER (F–O–R) SWITCH

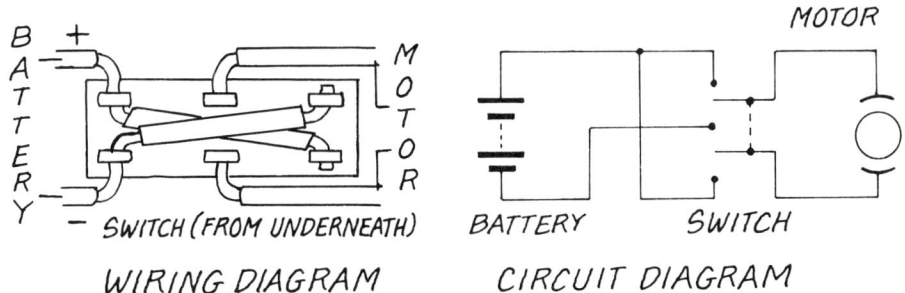

BATTERY
MOTOR
SWITCH (FROM UNDERNEATH)
WIRING DIAGRAM

MOTOR
BATTERY
SWITCH
CIRCUIT DIAGRAM

SOLENOIDS   ELECTROMAGNETISM

SOLENOIDS

When electricity flows through a wire a **magnetic field** is created around the wire. If you hold a compass near to a wire and switch the power on and off, the compass needle will move.

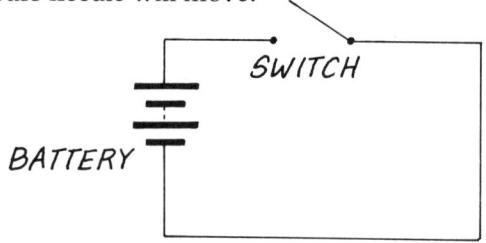

SWITCH

BATTERY

COMPASS

You will also find that the greater the current (try different batteries) the more powerful is the magnetic field.

This magnetism is not very strong. To make a more powerful **electromagnet** (solenoid), wind several turns of insulated copper wire (shellac, cotton or plastic coated) round a nail or soft iron rod.

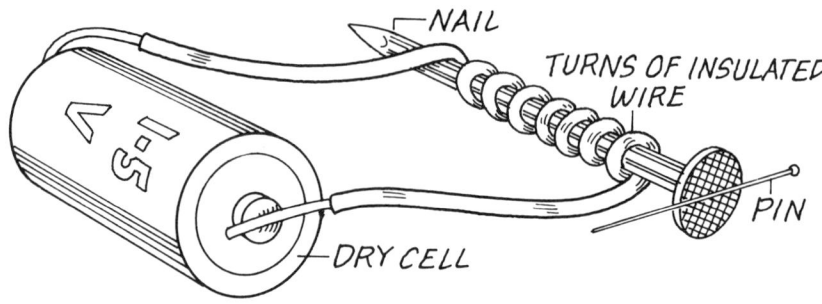

NAIL

TURNS OF INSULATED WIRE

PIN

DRY CELL

Increasing the number of turns increases the power of the magnet. However, the longer the wire the greater will be its resistance. Remember the greater the resistance the weaker the current and, therefore, the weaker the magnetic field!

A USEFUL SOLENOID

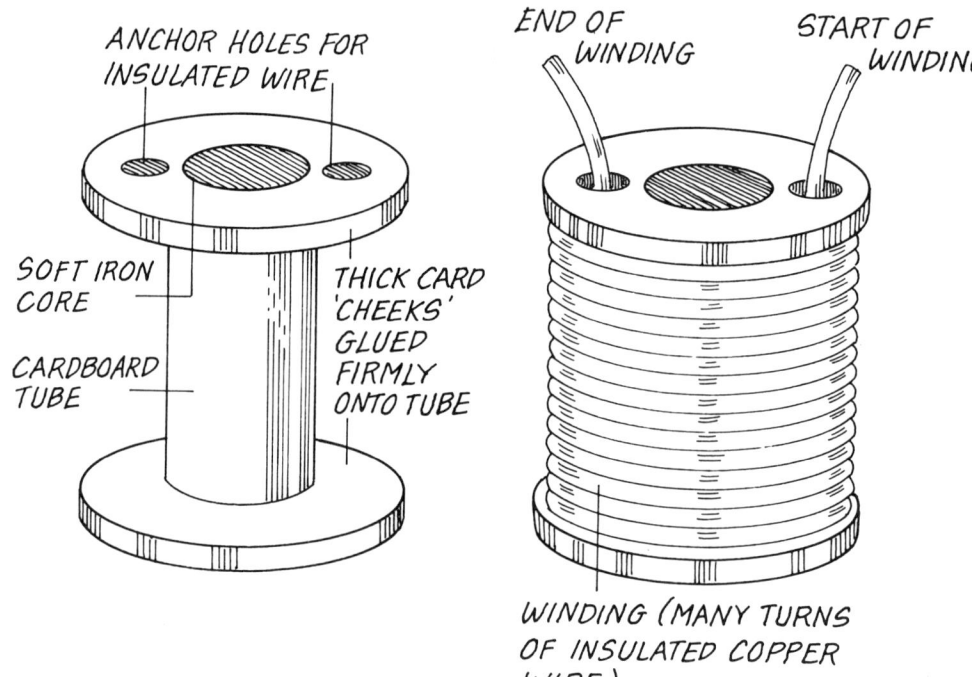

ANCHOR HOLES FOR INSULATED WIRE

END OF WINDING

START OF WINDING

SOFT IRON CORE

THICK CARD 'CHEEKS' GLUED FIRMLY ONTO TUBE

CARDBOARD TUBE

WINDING (MANY TURNS OF INSULATED COPPER WIRE)

Make a core for the winding from a piece of soft iron rod. If you prefer, bundle together enough soft iron wires to tightly pack the central hole of the cardboard tube. The core will work better if the wires are insulated from each other.

WARNING: A solenoid will soon run down your battery, so don't use it more than necessary.

If you want to know more about solenoids, find out about relays, bells and **buzzers**.

ELECTRONICS

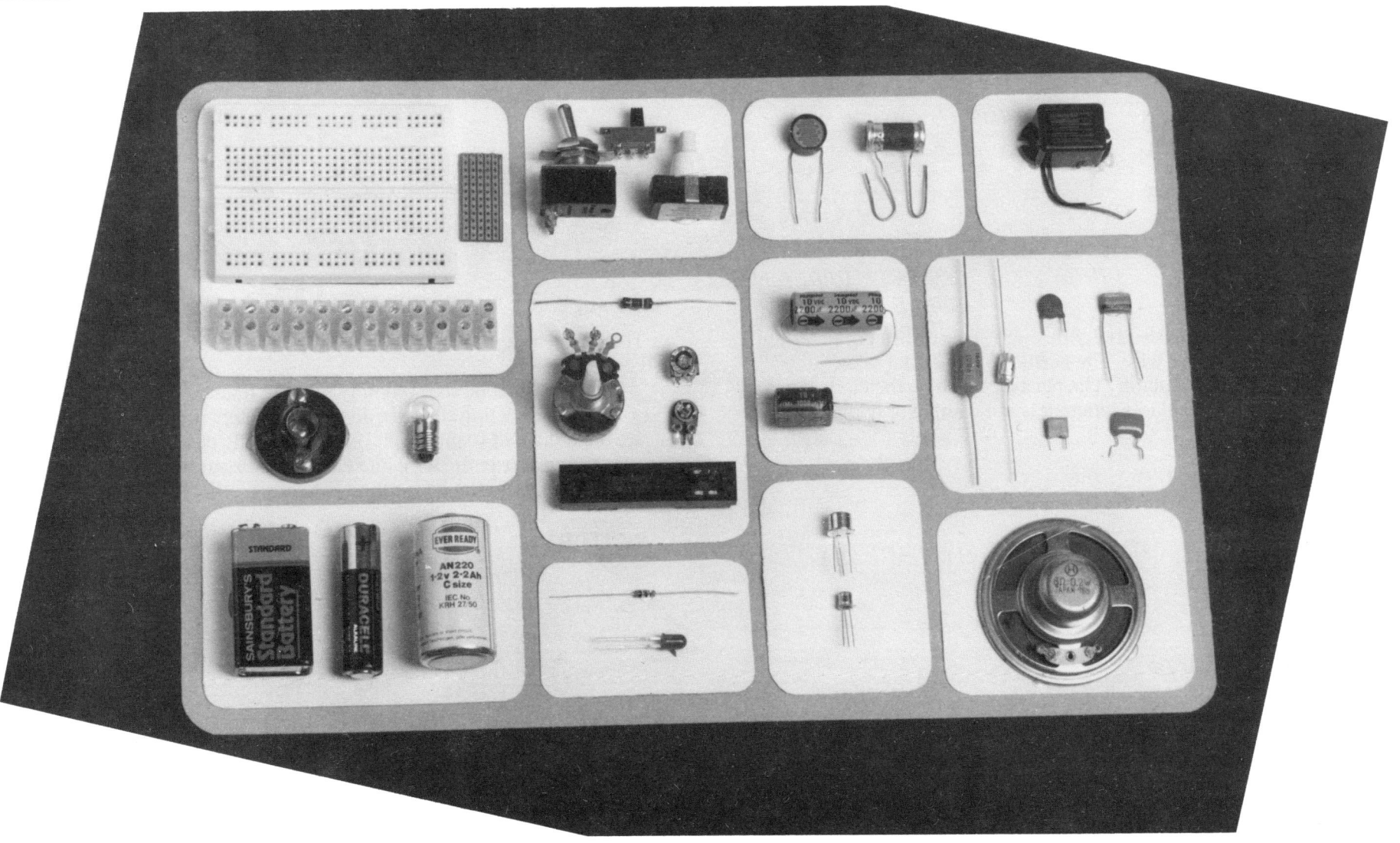

From left to right. Column 1: circuit boards, lamp socket and lamp, cells and battery. Column 2: switches, resistors, diodes. Column 3; sensors, electrolytic capacitors, transistors. Column 4: buzzer, capacitors, loudspeaker. **Can you find out the circuit symbol which goes with each of these components?**

ELECTRONICS

INTRODUCING THE FIRST COMPONENTS

CONNECTING WIRE

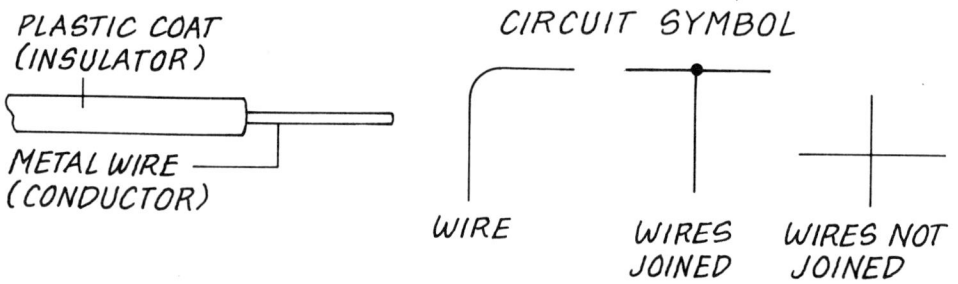

PLASTIC COAT (INSULATOR)

METAL WIRE (CONDUCTOR)

CIRCUIT SYMBOL

WIRE

WIRES JOINED

WIRES NOT JOINED

Only the metal part allows the electricity through. This is the **conductor** and is of tin-coated (tinned) copper wire. The plastic coat does not allow electricity through. It is an **insulator**. The plastic coat stops the electricity 'leaking' (short-circuiting) to places where it is not wanted or could cause damage.

SIGNAL LAMP

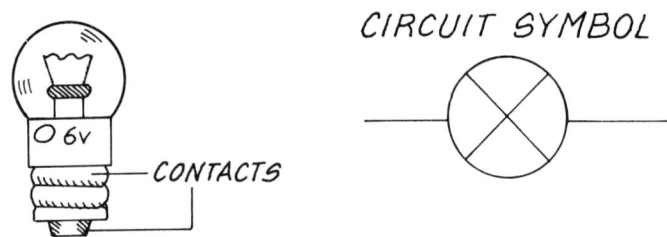

O 6v

CONTACTS

CIRCUIT SYMBOL

This has a fine wire (**filament**) inside made from the metal tungsten. This wire conducts electricity and acts as a resistance. If the flow of current is great enough, the filament glows. The greater the flow of electricity, the hotter the filament, the more brightly it glows: red – orange – yellow – white hot. If the current is too great the filament melts and breaks the circuit. The bulb is now 'blown' or 'fused' and has to be thrown away.

LAMPHOLDER

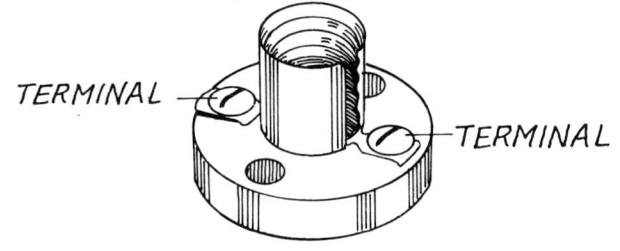

TERMINAL

TERMINAL

This device is for holding a bulb and providing screw terminals for easier connection.

The bulb must be screwed fully into the holder.

When connecting the wire do *not* completely unscrew terminals as they are often difficult to put back. Undo them only enough to wrap the bare wire ends round. Wrap the wire in a clockwise direction so that tightening the screw helps wind the wire round the screw, otherwise it often comes undone.

SWITCH

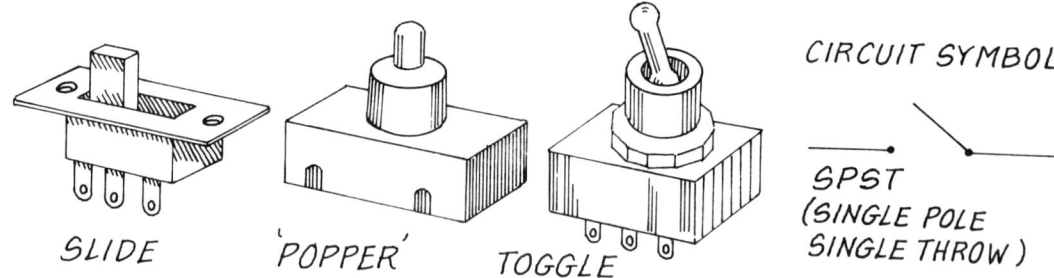

CIRCUIT SYMBOL

SLIDE

'POPPER'

TOGGLE

SPST (SINGLE POLE SINGLE THROW)

A switch is simply a break in the circuit which can be opened or closed. This could be done just by having a wire which is connected when the circuit is on and not connected when it is off. The kind of switch shown above is called a single pole single throw switch (spst). This means it only works on one wire and needs one movement to change from on to off and vice versa.

THE BREADBOARD

The **breadboard** is specially designed for trying out electronic circuits without having to solder. This means it is quick to use and cheap because with a little care components can be used again and again.

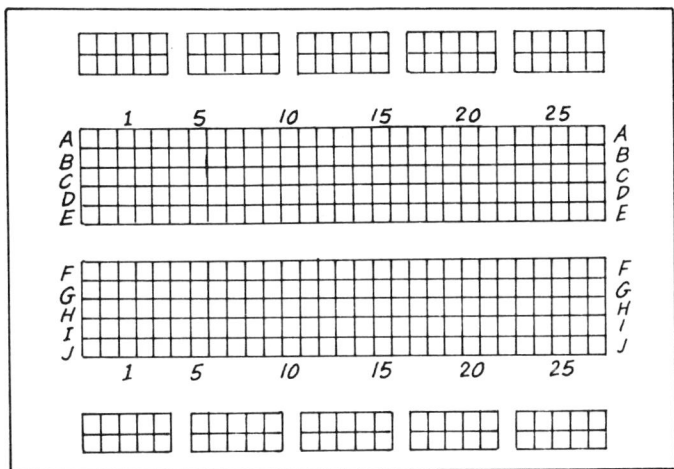

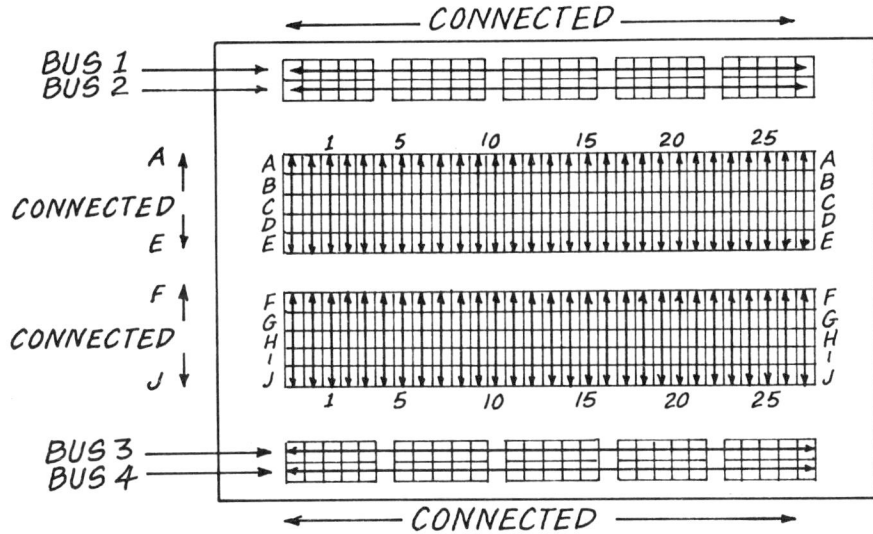

There are four buses (conducting channels) – each of these is a continuous connection from the extreme left hole to the extreme right hole. The centre patches of connectors are divided into two sets. The top set 'abcde' are connected vertically. There is no connection across the central island. The bottom set 'fghij' are also connected vertically.

It is important that single core wire is used to make connections on the board. On no account should stranded be used. The ends of the wire should be straight and pushed into the holes very firmly.

TERMINAL BLOCK CIRCUITS

This is a cheap way of making more permanent circuits with very little soldering. The **terminal block** is a series of twelve **insulated** connectors with a clamping screw at each end of every conductor.

The main problems arise from not getting the wires firmly clamped down by the screw or by clamping onto the plastic coating on the wire instead of on the metal core.

Most of the circuits are given using breadboard and terminal block layouts.

STRIPBOARD

A **stripboard** is for permanent circuits and requires you to be good at soldering. There are no wiring diagrams for stripboard. It is assumed that if you can use stripboard you can also work out your own wiring diagram!

THE BATTERY

CIRCUIT SYMBOL

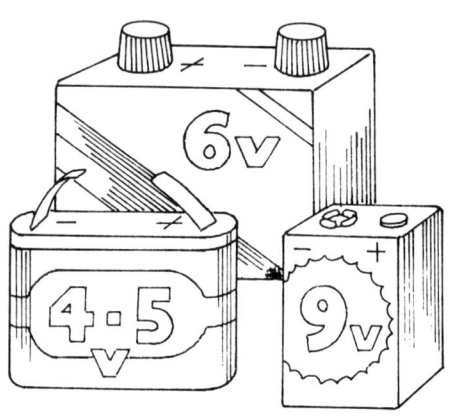

The long line shows the positive (+) connection. The short line shows the negative (−) connection. The voltage is usually written next to the battery e.g.

There is a separate information sheet on batteries (Energy & Control 4).

THE CIRCUIT

Circuit is the name given to the path (or paths) travelled by the electrical current in going from the positive to the negative side of the battery. It is like the 400 m hurdles in athletics: the start is the battery positive, the finish is the battery negative, the runners are the current and the hurdles are the **components**, like the bulb in a circuit, where the current has to do extra work. The circuit must be complete for the current to flow; if the circuit is broken the current stops (see Switch).

Can you name all the components in this circuit?

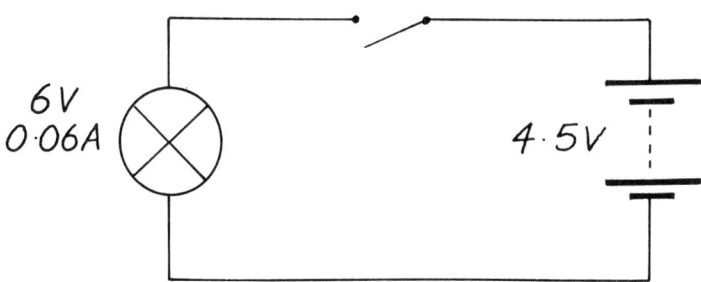

SINGLE LAMP CIRCUIT

Here then is the first practical circuit. There are wiring diagrams shown below using either 2 A terminal blocks or a small **breadboard** or prototyping board. (For details of the breadboard see Electronics 2.) Connect up the circuit as shown. When you make the wire switch complete the circuit, the bulb should light. It will be brighter with the 6 V than with the 4.5 V battery. If you take the switch wire out, the bulb will go off even though one side is still connected to the battery. Can you explain why?

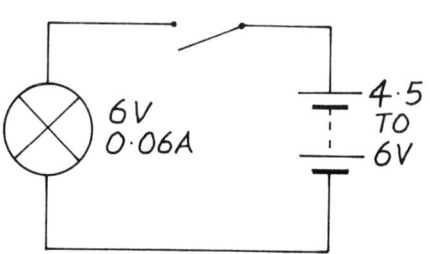

WIRING DIAGRAMS

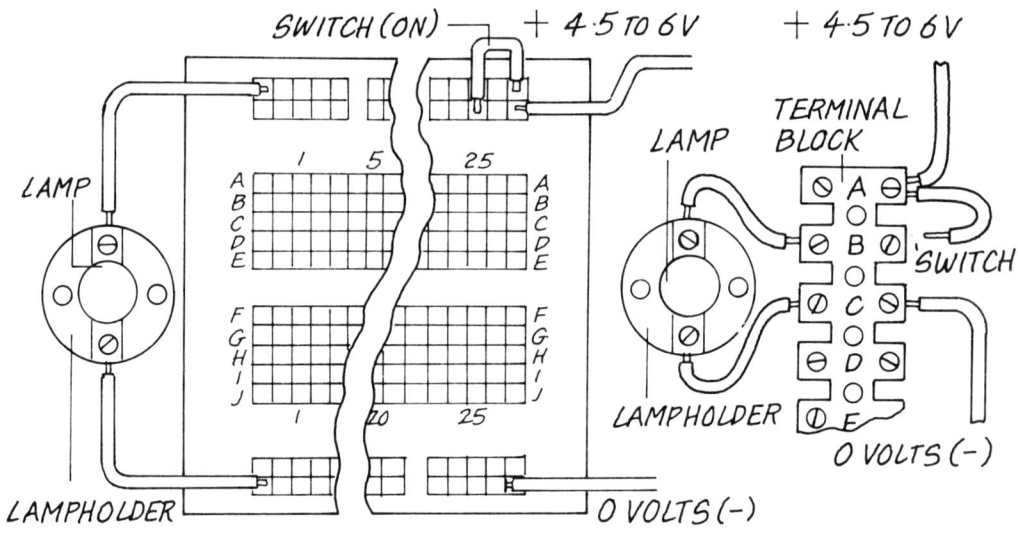

(The breadboard is shown with the middle cut out to save space.)

LAMPS IN SERIES

Series means in a chain or one after the other. When you have set up this circuit and 'switched' on, you should notice that both lamps are dimmer than in the single lamp circuit. If either of the lamps is unscrewed, what happens to the other one? Can you explain this? Can you change the circuit to put another lamp in the series? What do you notice this time? How many lamps can you have in the chain before the filaments stop glowing? Why do the filaments stop glowing? Is current still flowing when the filaments are dark?

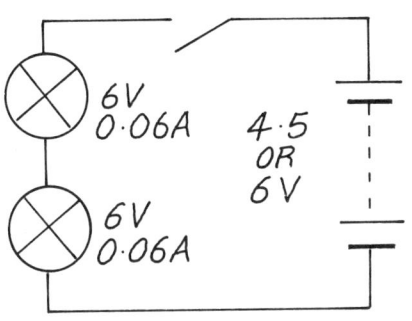

LAMPS IN PARALLEL

Parallel means side by side. This means the current has two pathways by which it can complete the circuit rather than the one path given by the series circuit. What do you notice about the brightness of the lamps this time? If either of the lamps is unscrewed what happens to the other one? Can you explain this? Can you change the circuit to put another lamp in parallel to the others? What do you notice this time? How is this different from the series circuit? Unscrew each bulb in turn in this circuit and note what happens in each case.

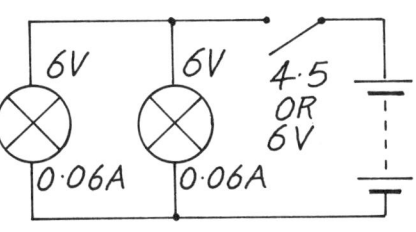

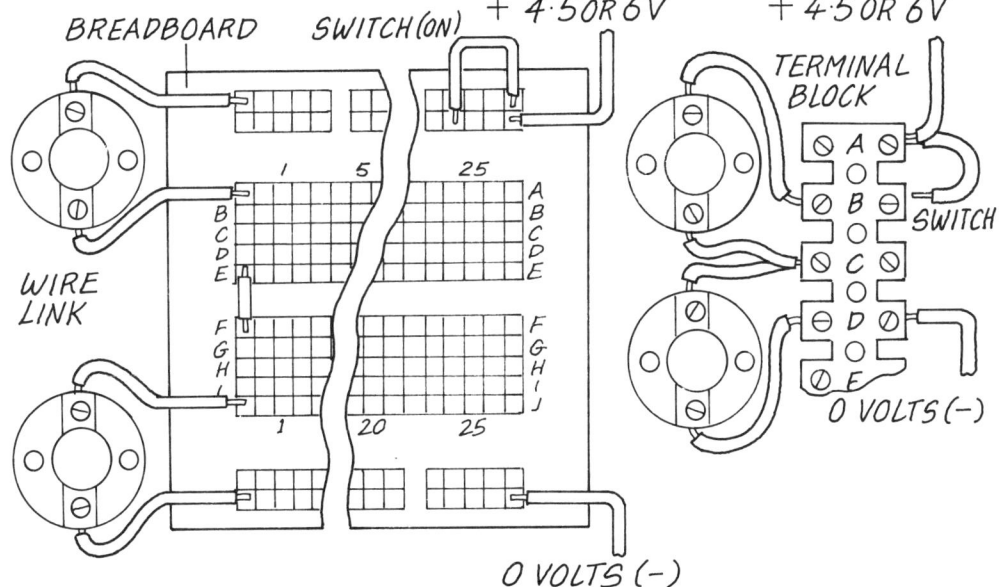

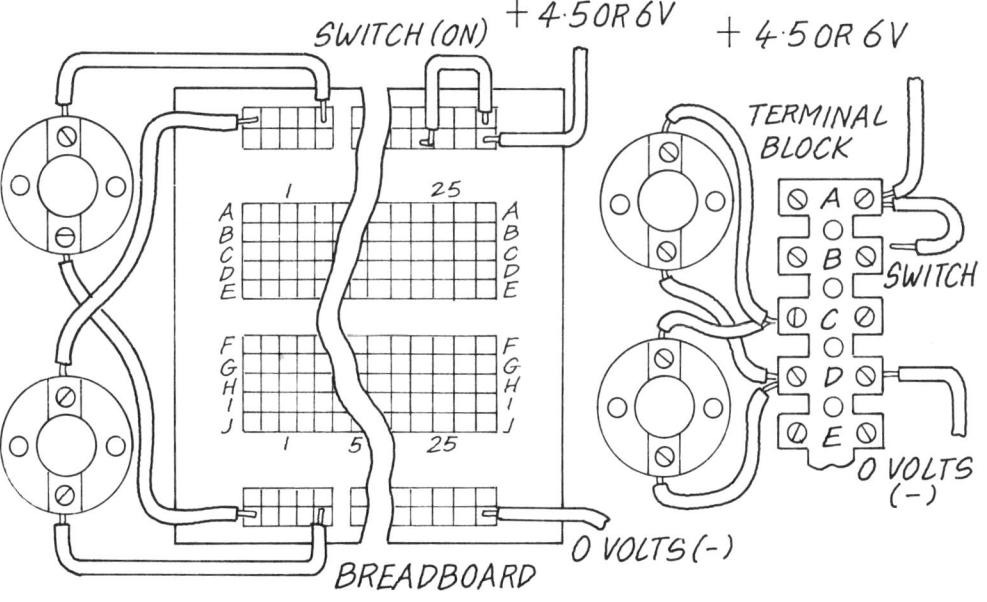

(The breadboards are drawn with the middle part missing to save space.)

THE RESISTOR

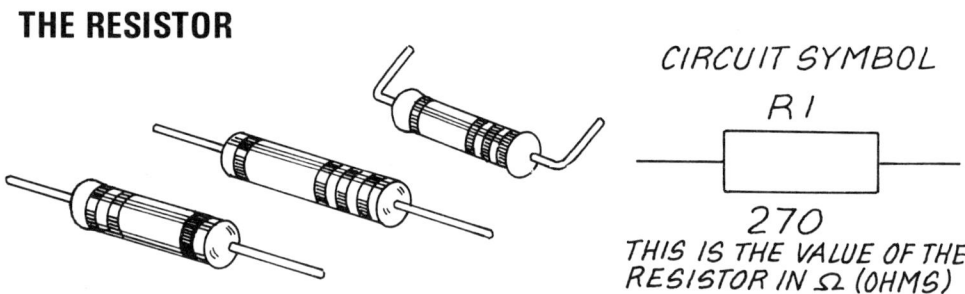

CIRCUIT SYMBOL

R1

270
THIS IS THE VALUE OF THE
RESISTOR IN Ω (OHMS)

A **resistor** is a **device** which reduces the current flowing in a circuit. The greater the resistance the smaller the current. Electrical resistance is measured in ohms, usually written as Ω. Think of water flowing in a garden hose: if you squash the tube with your foot, the flow of water is reduced, perhaps even stopped. The squashed tube is acting as a resistance to the flow of water; the more the tube is squashed the greater the resistance and the smaller the flow of water.

The physical size of a resistor has nothing usually to do with its resistance. The size has to do with the maximum electrical power it can handle, measured in watts. In most of the circuits we use, the resistors are rated at either 0.25 ($\frac{1}{4}$) watt or 0.5 ($\frac{1}{2}$) watt.

THE RESISTOR COLOUR CODE

The resistor normally has four coloured bands round its body. These are to tell you the value of the resistor in ohms and how accurate this value is.

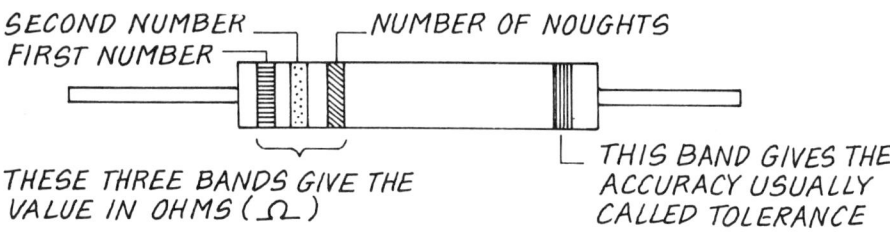

SECOND NUMBER
FIRST NUMBER   NUMBER OF NOUGHTS

THESE THREE BANDS GIVE THE
VALUE IN OHMS (Ω)

THIS BAND GIVES THE
ACCURACY USUALLY
CALLED TOLERANCE

Each colour of the three value bands is equal to a number. The table on the right shows the number for each colour.

The fourth band is gold or silver on all common resistors. It gives the tolerance (expected accuracy) of the resistance. Gold is ±5 % accurate. Silver is ±10 % accurate. This means a 100 Ω resistor with a silver tolerance band is between 90 Ω and 110 Ω but a 100 000 Ω resistor of the same tolerance will be between 90 000 Ω and 110 000 Ω!

Band colour	Number
black	0
brown	1
red	2
orange	3
yellow	4
green	5
blue	6
violet	7
grey	8
white	9

READING THE CODE

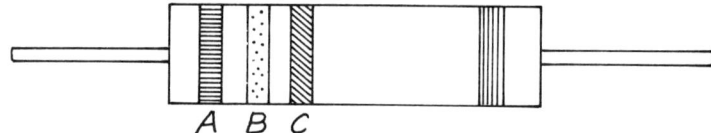

A B C

To read the value of a resistor start with the first band (**A**). This gives the first number. The second band (**B**) gives the second number. The third band gives the number of noughts which follow the first two numbers. For example suppose **A** = yellow, **B** = violet and **C** = orange; our resistor would be 47 000 Ω yellow = 4, violet = 7 and orange = three noughts = 000. 47 000 Ω is a bit longwinded so we shorten it to 47 kΩ.

kΩ is short for kilohm = 1000 Ω = 1 k.
MΩ is short for megohm = 1 000 000 Ω = 1 M.

(REMEMBER: kilo = thousand and mega = million.)

RESISTOR IN SERIES   RESISTOR IN PARALLEL

THE RESISTOR IN SERIES

If resistor R1 is 100 Ω, what effect does this have on the brightness of the lamp? Try larger values of R1. How do these affect the lamp? It is important to realise that although the lamp seems not to be working at all, there is current flowing through the circuit. This current is too small to make the filament glow even red hot.

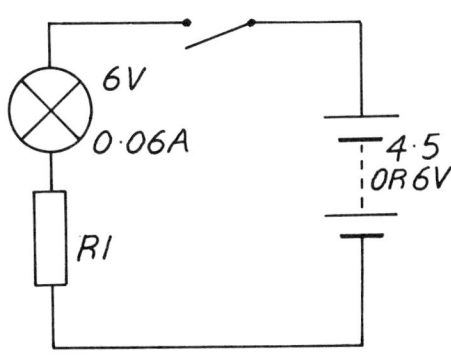

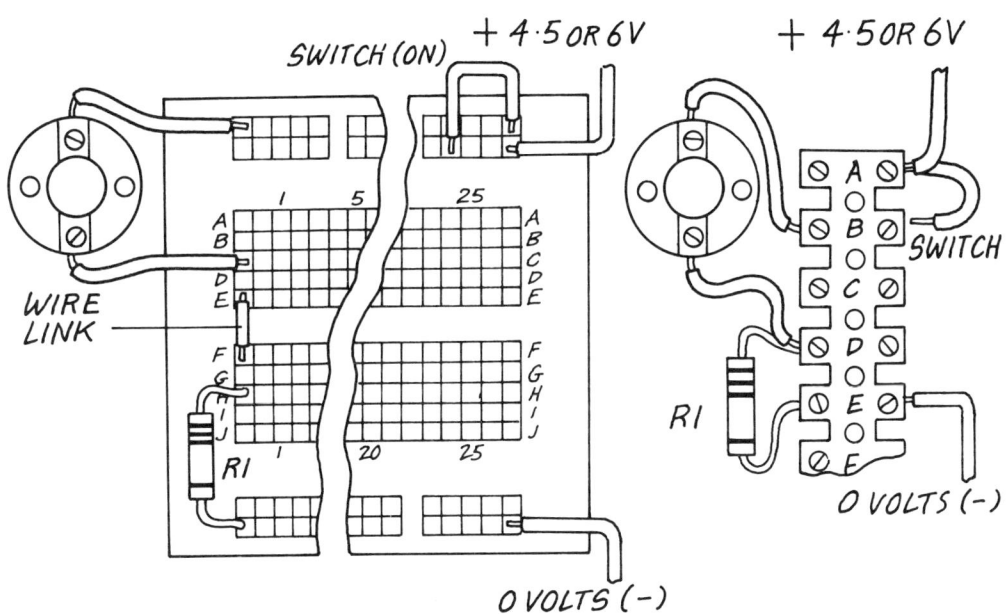

A PARALLEL RESISTOR

L1 is brighter than L2. Why? Try disconnecting one end of R1 then touching it in and out of the circuit. What do you notice? Unscrew L1 and everything goes off. Unscrew L2 and L1 goes dimmer but stays on. Can you explain this? After going through L1, there are two pathways for the current to flow. The resistance of L2 and R1 together in parallel is less than the resistance of L2 or R1 separately. This means more current flows through L1 so it is brighter. However because some current flows through R1, less flows through L2 so it glows less brightly.

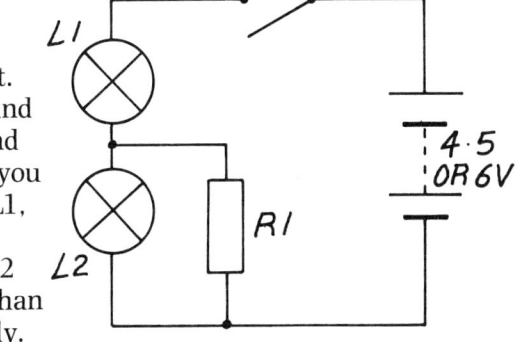

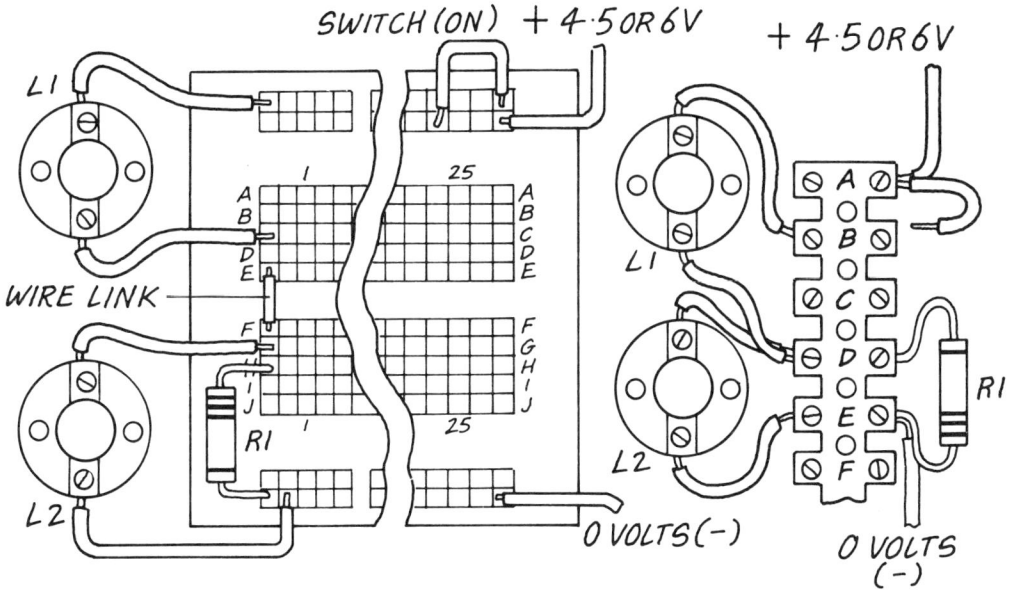

(NOTE: the breadboards are drawn with the middle cut out to save space.)

THE VARIABLE RESISTOR (POTENTIOMETER)

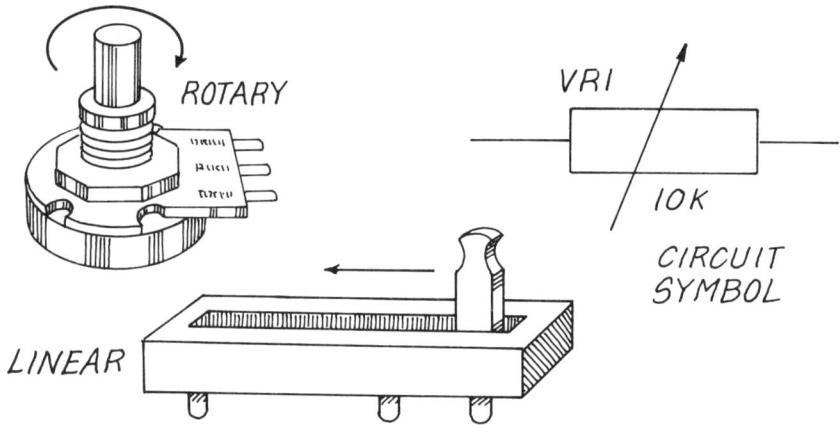

Although very different in appearance from the fixed resistor, the **variable resistor** (or **potentiometer**) does the same sort of job. It reduces the current. The variable resistor in addition can vary the amount by which the current can be reduced from maximum to minimum. The arrows in the diagram show how to change the resistance. The variable resistor has three connections – the middle connector is the one that must be connected and one of the outer two if the resistance is to be varied.

THE PRESET RESISTOR

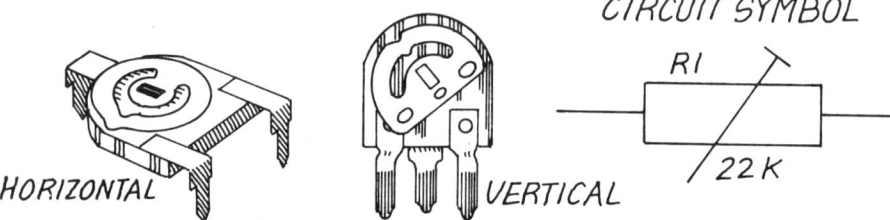

The **preset** is a form of variable resistor which can be set with a screwdriver to the required value. They are used mainly when there is not much need to keep changing the preset value.

THE VARIABLE RESISTOR IN SERIES

Notice that this circuit is the same as the 'series resistor' on sheet 6, with VR1 in the place of R1. What happens when you change the resistance of VR1? Why? You could also try replacing R1 with VR1 in the 'resistor in parallel' on sheet 6 then observe what happens to L1 and L2 as you change the value of the resistance from 0 to 470 Ω. Remember that current is still flowing even when the lamps remain dark, as long as the circuit is complete.

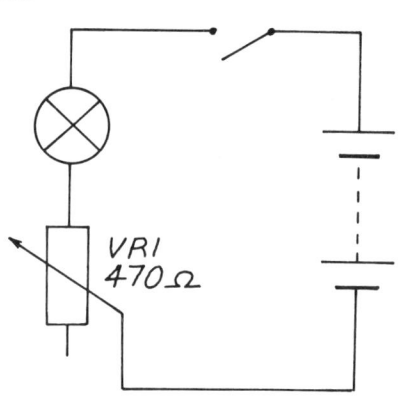

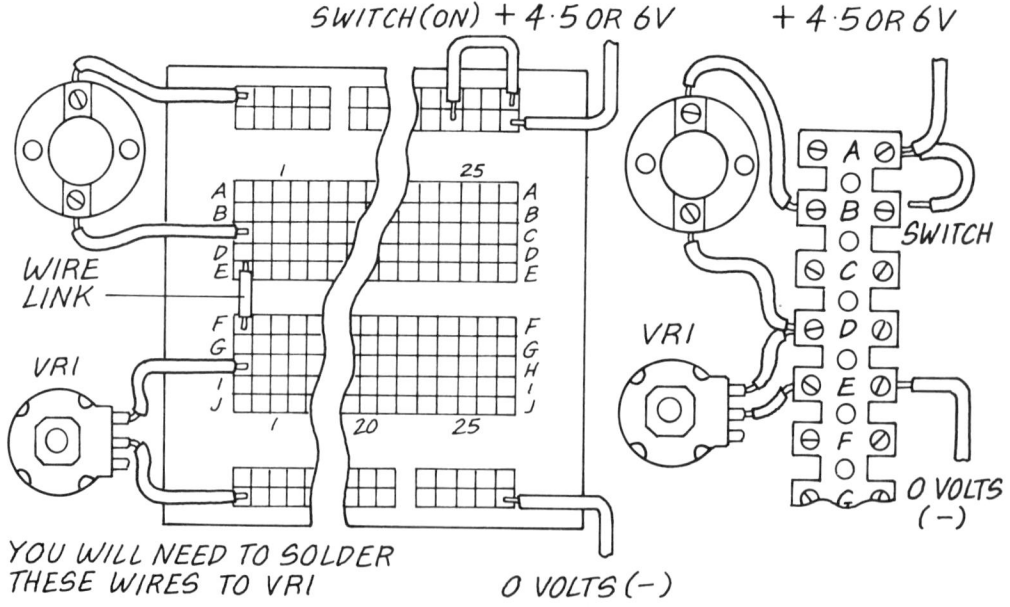

YOU WILL NEED TO SOLDER
THESE WIRES TO VR1

DIODES – GENERAL

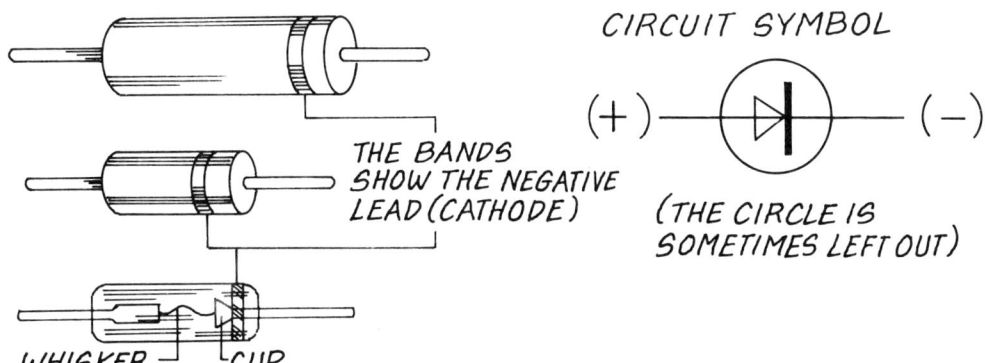

CIRCUIT SYMBOL

THE BANDS
SHOW THE NEGATIVE
LEAD (CATHODE)

(+) —▷|— (–)

(THE CIRCLE IS
SOMETIMES LEFT OUT)

WHISKER — CUP

A **diode** will allow current to flow through in one direction only. Think of a turnstile at a sports ground and the spectators only able to go through in one direction, or think of a one-way street. Diodes are used whenever the current is required to go only in the one direction. Some diodes need to be strong enough to take large currents. They are used to change **alternating currents (a.c.)** (where the current changes direction rapidly from forwards to backwards to forwards to backwards) into **direct current (d.c.)** (where the current stays fixed in one direction). These diodes are called **rectifiers** and are used in power packs.

Diodes are also used to protect equipment from damage caused by connecting batteries or wires the wrong way round. Can you work out a circuit to do this?

LIGHT-EMITTING DIODES (LEDs)

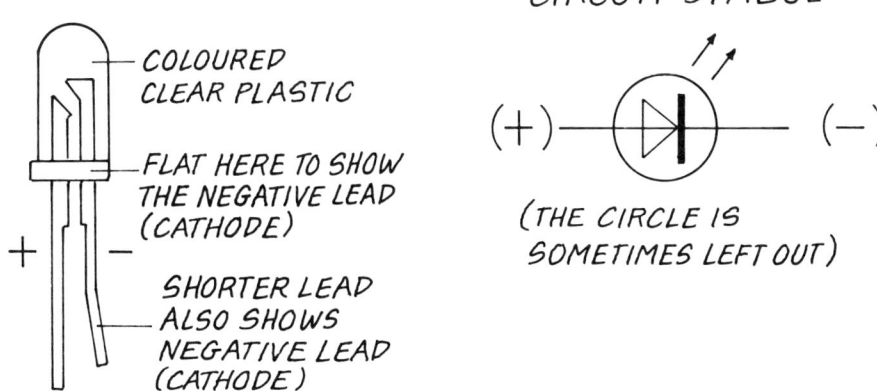

CIRCUIT SYMBOL

COLOURED
CLEAR PLASTIC

FLAT HERE TO SHOW
THE NEGATIVE LEAD
(CATHODE)

SHORTER LEAD
ALSO SHOWS
NEGATIVE LEAD
(CATHODE)

(+) —▷|— (–)

(THE CIRCLE IS
SOMETIMES LEFT OUT)

Here is another type of diode which is very useful. Like all diodes, it will only allow the current to flow through in one direction. Unlike the other diodes, it gives off light when there is a current flowing through. **Light-emitting diodes (LEDs)** are usually found in three colours – red, green and yellow. They are usually round but can be other shapes.

An LED is easily damaged if it is connected directly across the battery. The current is much too strong and burns it out. Always use a resistor, usually 680 Ω, in series with the LED to reduce the current and prevent damage, as shown in the circuit diagram.

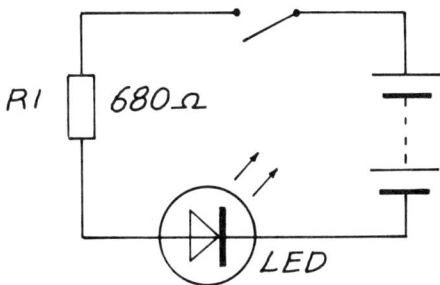

R1   680 Ω

LED

A DIODE CIRCUIT

Set up the circuit shown on the right. Make sure the negative lead of the diode (the end with the band round) goes to the 0 V (battery negative) side of the circuit. Switch on and the lamp should light. Switch off and turn the diode round (so the end with the band is nearest the lamp). When you switch on this time the lamp will stay off. This shows that the diode will only allow current to flow through it in one direction.

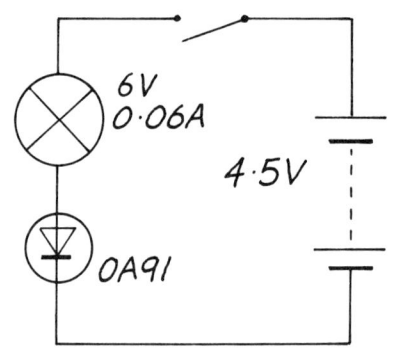

6V
0·06A

4·5V

0A91

AN LED CIRCUIT

When you set up this circuit make sure that you connect the LED the right way round (see Electronics 8). When you switch the circuit on, the LED should light up. (If it doesn't, turn the LED round so the leads are connected up the opposite way.) What happens if the LED is connected up the other way round? Would the same happen if we used a lamp instead of the LED and resistor?

CAUTION: Do not leave out the resistor or you will damage the LED.

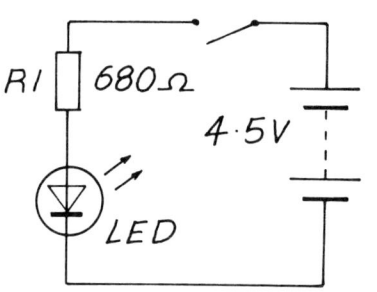

R1   680 Ω

4·5V

LED

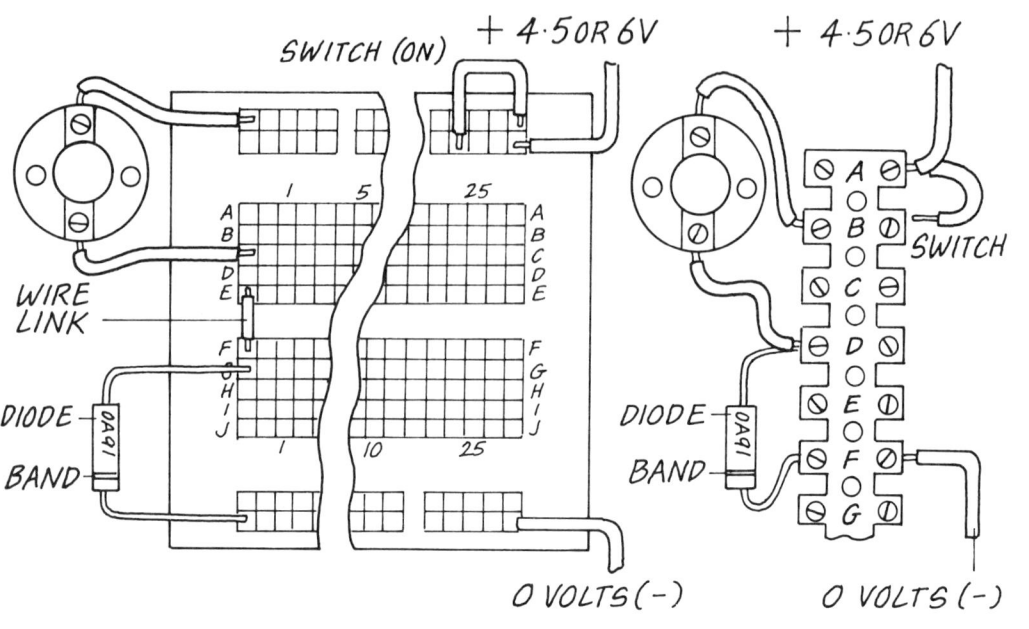

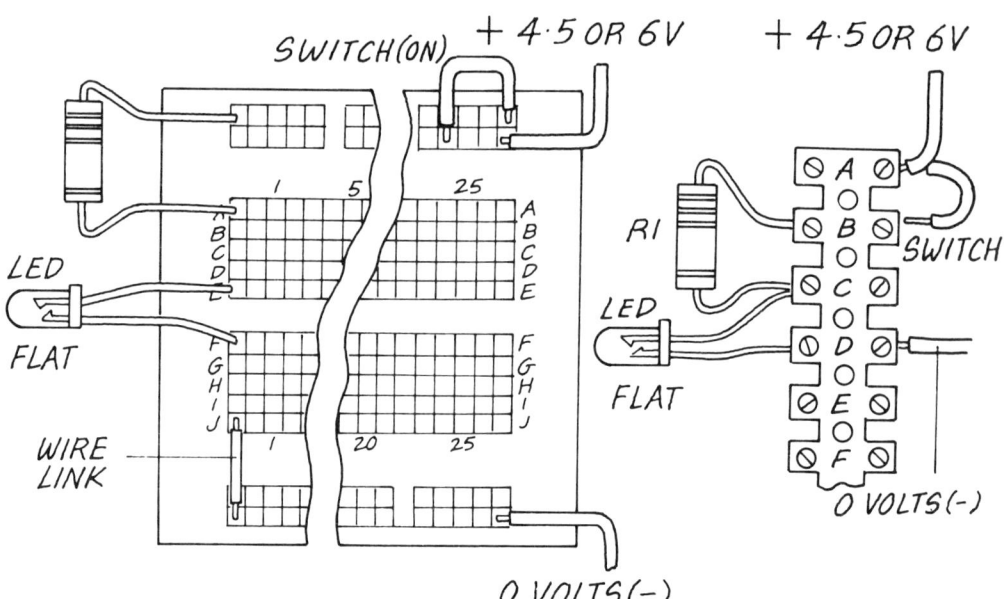

THE NPN TRANSISTOR

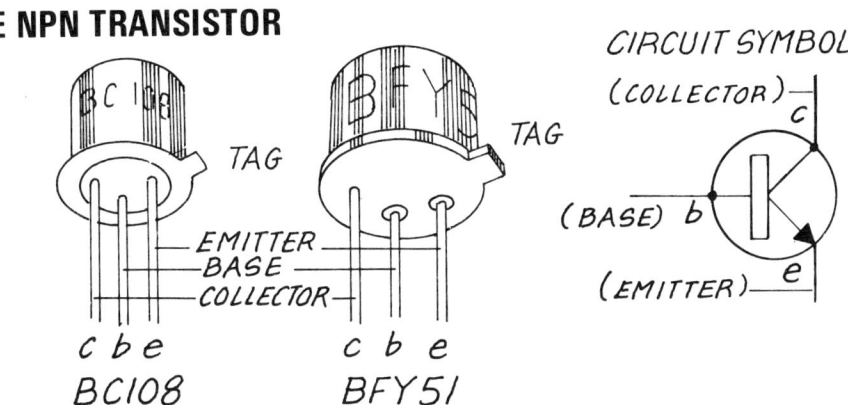

The **transistor** has three leads; each one has a name. The leads are called the **collector**, the **base** and the **emitter**. An NPN transistor is one that must be connected so that the collector is to the positive side, the base must also be positive and the emitter is to the negative or 0 V side of the circuit. A transistor will not allow a current to flow between the collector and the emitter unless there is a small current present at the base. The transistor is like a switch in the off position when there is no current, or too little current. If there is a large enough base current, then a much larger current can flow between the collector and the emitter. The transistor is now like a switch in the on position. The transistor acts as a current-controlled switch with the base as the on/off toggle.

AS SWITCH

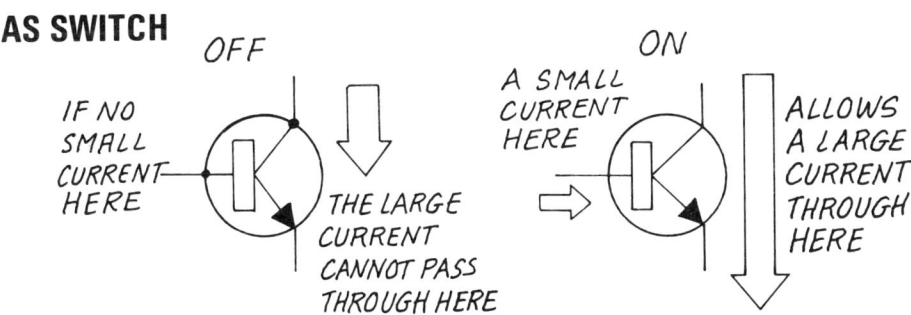

AS AMPLIFIER

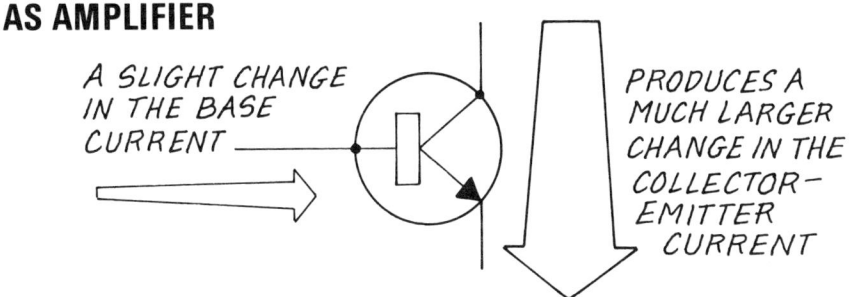

A small increase in the base current will produce a much larger increase in the collector–emitter current. The transistor acts as a current **amplifier**. There is a maximum current that the transistor can safely conduct without being damaged or destroyed. If too high a current flows, its resistance falls so more current flows. The transistor gets hotter and hotter and overheats. This is called thermal runaway and destroys the transistor. The BFY51 is a more powerful device than the BC108 and can take about ten times the collector current without damage.

DARLINGTON PAIR

For some circuits we connect two transistors together to increase the sensitivity and gain (current **amplification**). The arrangement shown here is called a Darlington pair amplifier.

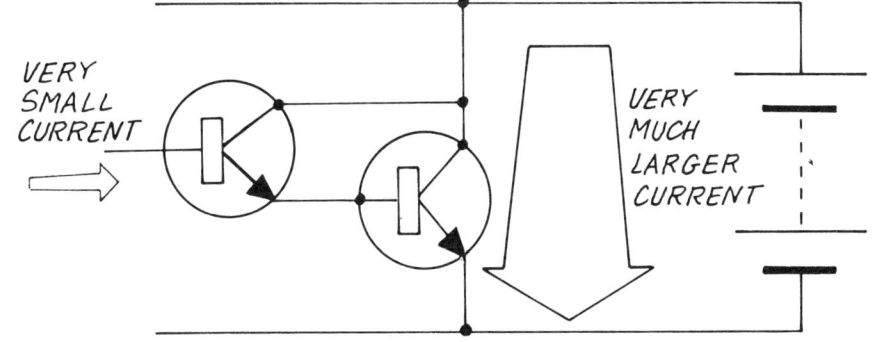

THE TRANSISTOR SWITCH CIRCUIT

THE TRANSISTOR AS SWITCH

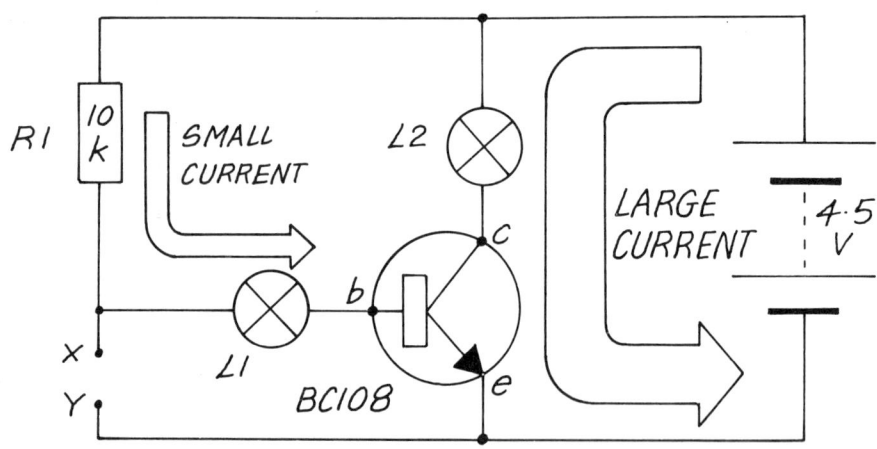

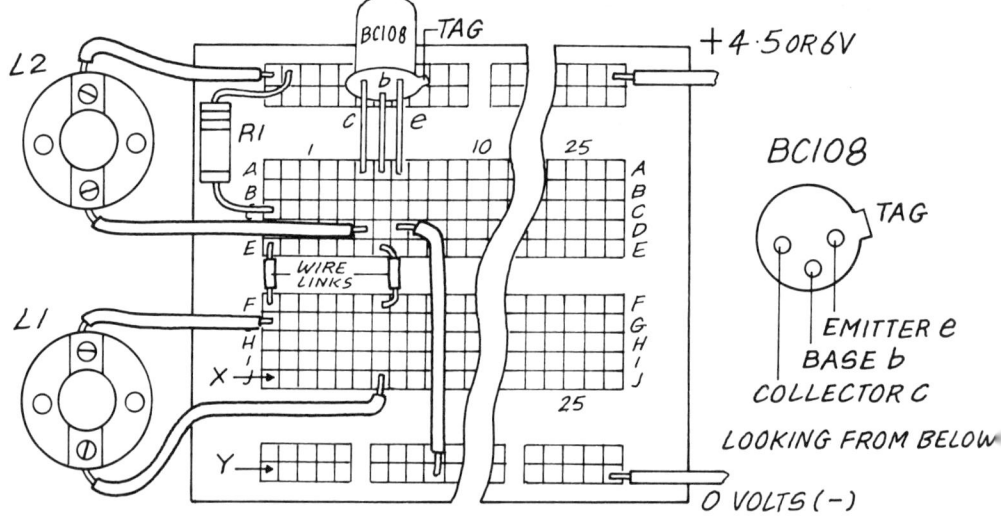

When the battery is connected lamp L2 lights up but lamp L1 remains dark. L1, though dark, is not idle! If you unscrew L1, L2 goes off. Why? Do you remember the circuit with the lamp and the series resistor? (See Electronics 6.) This showed that any resistor much bigger than 100 Ω reduced the current so much it was too small to make the lamp filament glow. Here we have a 10 k resistor in series with lamp L1 so the current is a hundred times smaller – no wonder the lamp doesn't glow! This very small current flowing through the base of the transistor is enough to switch the transistor on. This allows a large current to flow through the collector–emitter lighting up lamp L2. Unscrewing L1 stops the flow of the base current which switches the transistor off and lamp L2 goes dark. Now try connecting a wire from **X** to **Y**. L2 should go dark. Why? This is because the current always goes by the easiest path. It is much easier for the current to flow straight back to the negative (0 V) side of the battery than to go through the lamp L1 and the base of the transistor. The transistor is now off so there is no collector–emitter current and L2 goes out. We will be making use of this in the sensor circuits in Electronics 13.

The transistor is shown twisted over so you can identify the three connecting leads more easily. These three leads are labelled **c**, collector, **b**, base, and **e**, emitter, and must be connected correctly as shown or the transistor could be destroyed. Check your circuit before connecting the battery.

The transistor will need extension leads of single core insulated wire. These should be carefully soldered and the sleeving pushed over the part of the lead nearest the transistor. This is to prevent accidental short-circuits. (See Electronics 18.)

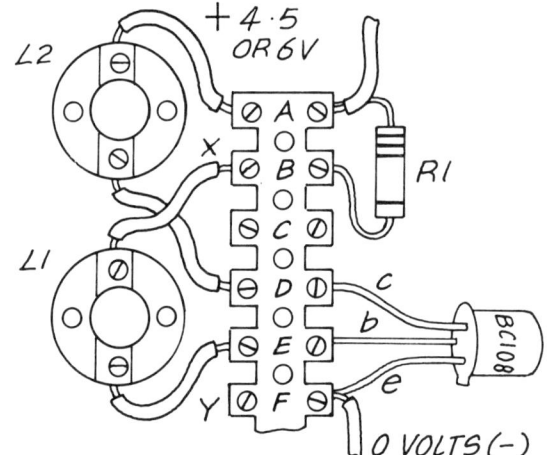

LIGHT-DEPENDENT RESISTOR (LDR)

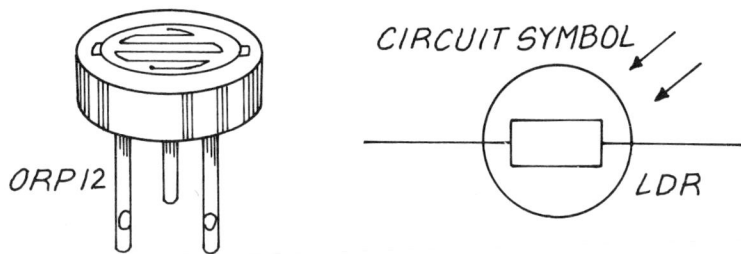

CIRCUIT SYMBOL

ORP12

LDR

The ORP12 is a **light-dependent resistor (LDR)**. It is a special form of resistor which uses a material called cadmium sulphide which has a low resistance when light is shining on it and a very high resistance when it is dark. In strong light the ORP12 has a resistance of as little as 150 Ω while its resistance in the dark is in the region of 10 M.

THERMISTOR

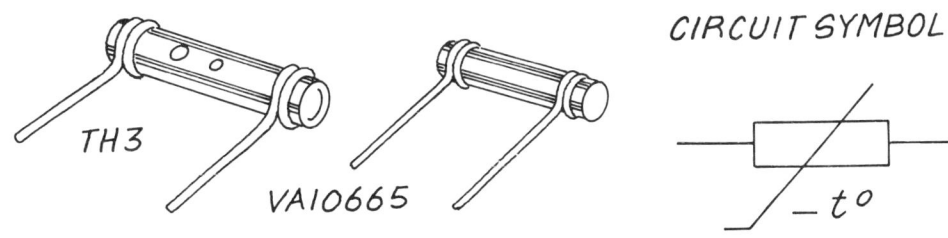

TH3

VA10665

CIRCUIT SYMBOL

$-t^o$

The **thermistor** is a device whose resistance falls as its temperature rises and becomes greater as its temperature falls. E.g. the VA1066S has a resistance of 4.7 k at 25 °C and a resistance of 200 Ω at 150 °C. The TH3 has a resistance of 380 Ω at 25 °C and a resistance of 28 Ω when hot.

WET/DRY SENSORS

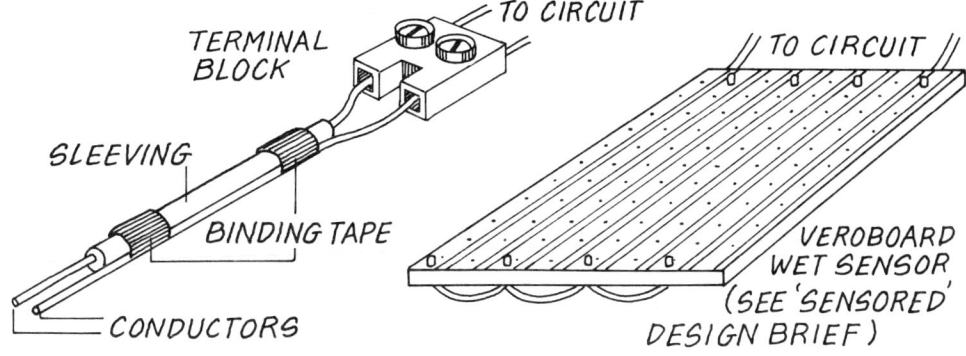

TO CIRCUIT

TERMINAL BLOCK

TO CIRCUIT

SLEEVING

BINDING TAPE

CONDUCTORS

VEROBOARD WET SENSOR (SEE 'SENSORED' DESIGN BRIEF)

These are sensors you can make for yourself. They rely on the fact that water, unless absolutely pure, conducts electricity. By arranging two conductors to be near each other (about 1 mm apart) any drop of water which bridges the gap will allow a current to flow from one conductor to the other. This is again a change in resistance.

ALARMS – THE TRANSISTOR OSCILLATOR BUZZER

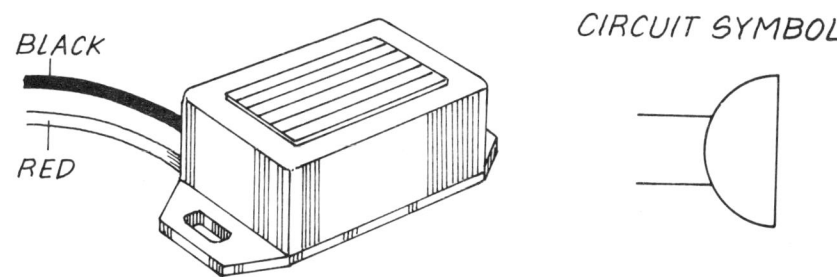

BLACK

RED

CIRCUIT SYMBOL

Sometimes it is more useful to have a sound alarm rather than the light of a lamp or LED. The transistor **oscillator buzzer** can be used instead of the lamp in the sensor circuits. Make sure the red lead goes to the battery positive side and the black goes to the collector.

THE TRANSISTOR SWITCH AS SENSOR

THE TRANSISTOR SWITCH AS SENSOR

This is the basic circuit. By putting various sensors at either X or Y and a 10 k variable resistor at the other, several different sensor circuits are possible. Listed below are the different ways you can use this circuit. The sensitivity of the sensor is adjusted by means of the 10 k variable.

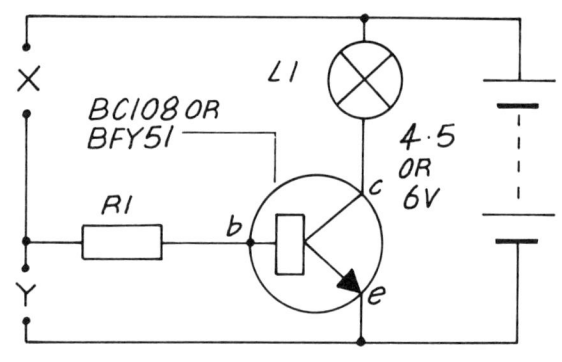

Type of sensor	X	Y	R1
dark sensor	10 k variable	ORP12	1 k
light sensor	ORP12	10 k variable	1 k
heat sensor	VA1066S/TH3	10 k variable	1 k
cold sensor	10 k variable	VA1066S/TH3	1 k
wet sensor	wires 1 mm apart	10 k variable	1 k
dry sensor BC108	10 k variable	wires 1 mm apart	39 k
dry sensor BFY51	10 k variable	wires 1 mm apart	15 k

EXAMPLES
E.G. LIGHT SENSOR

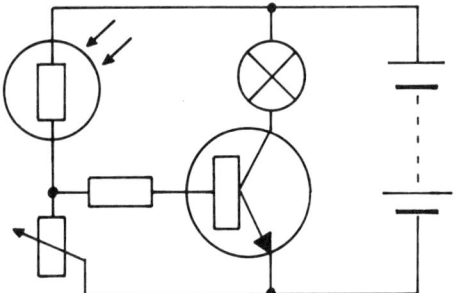

HEAT SENSOR

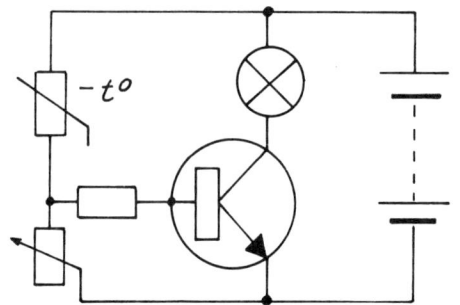

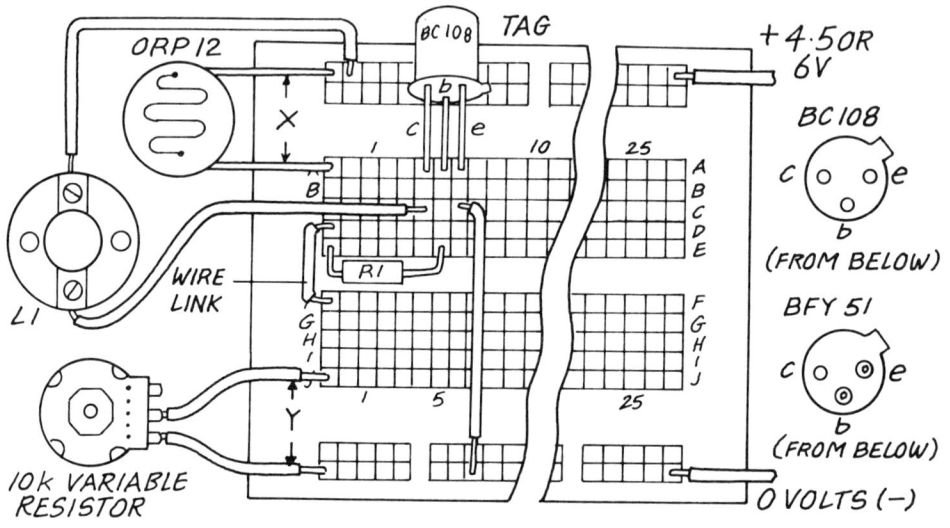

Here are the wiring diagrams for the breadboard and terminal block. They are shown wired as the light sensor. Extension leads will need to be soldered onto the 10 k variable resistor.

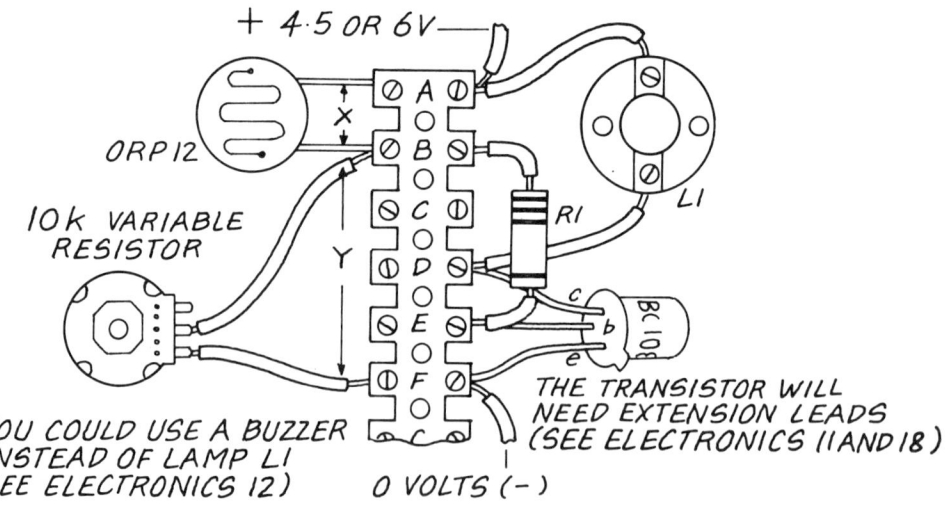

YOU COULD USE A BUZZER INSTEAD OF LAMP L1 (SEE ELECTRONICS 12)

THE TRANSISTOR WILL NEED EXTENSION LEADS (SEE ELECTRONICS 11 AND 18)

CAPACITORS – ELECTROLYTIC

Axial and radial refer to the way the leads come out of the can of the capacitor.

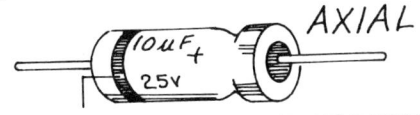

AXIAL

BLACK BAND TO MARK NEGATIVE

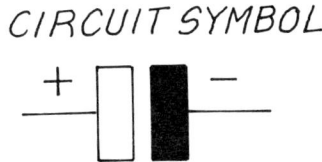

CIRCUIT SYMBOL

RADIAL

ARROW TO SHOW NEGATIVE

TYPES OF CAPACITOR

There are two main types of **capacitor**, electrolytic and non-electrolytic. The main difference between the two types is that the non-electrolytic can be connected with either lead to the positive or negative. The electrolytic *must* have the lead marked (−) connected to the negative and the lead marked (+) connected to the positive. All capacitors have a voltage marked on their cases. This is the maximum voltage that can safely be used with them. It is dangerous to exceed this voltage!

WHAT A CAPACITOR DOES

As can be seen from the circuit symbol, there is no pathway for current to flow directly through the capacitor. It acts as a barrier to direct current. (If you put a capacitor instead of the series resistor in the circuit in Electronics 6 the lamp does not light.) In direct current (d.c.) circuits the capacitor acts as a store of electricity. The greater the capacitance the more it can store. This ability is made use of in timing circuits. In these circuits the capacitor behaves like the cistern in a toilet. When full it holds its charge until it is

CAPACITORS – NON-ELECTROLYTIC

Ceramic, polystyrene and polyester refer to the materials from which they are made.

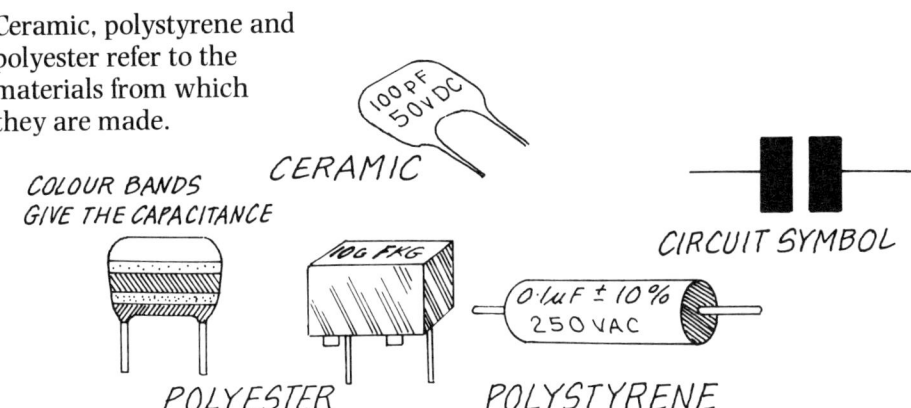

COLOUR BANDS GIVE THE CAPACITANCE

CERAMIC

CIRCUIT SYMBOL

POLYESTER   POLYSTYRENE

flushed (discharged). It then takes a certain time to fill up before it can be used again. With a cistern the time depends on the rate of flow of water (current) and how much water it holds when full (capacity). For a capacitor the recharge time depends on the current flowing into it and how much electricity it can hold. Capacitors are used as a reservoir (reserve) to even out small changes in the current. This is the 'smoothing' capacitor in power packs. In a.c. circuits the capacitor can effectively allow high frequency a.c. signals through while blocking low frequency signals and d.c. It is used as a filter in audio and radio circuits.

THE UNITS OF CAPACITANCE

Capacitance is measured in farads but these are about a million times too big for electronics. We use microfarads (μF) which are millionths of a farad, nanofarads (nF) and picofarads (pF). This is how it works.

1000 picofarads (pF) = 1 nanofarad (nF)
1000 nanofarads (nF) = 1 microfarad (μF)
1 000 000 picofarads (pF) = 1 microfarad (μF)

NB μ is the Greek letter 'mu'.

A MORE-SENSITIVE SWITCH

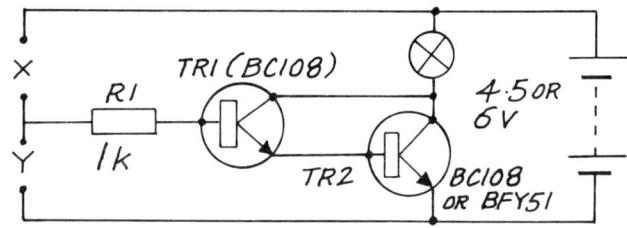

This circuit makes a much more sensitive switch than the single transistor version. Its sensitivity is due to the transistors being connected as a **Darlington pair**. Not only is the sensor circuit more sensitive, it is capable of switching on and off a small battery motor. To do this TR2 should be a BFY51 and the motor should require less than 1 amp (it could also switch a relay on and off). For the sensor circuit, the possible variations for **X** and **Y** are the same as shown in the table given in Electronics 13 with one exception: the dry sensor needs a 100 k variable resistor at **X** and the sensor at **Y**.

A TIME DELAY CIRCUIT

To make a simple time delay circuit, use a resistor at **X**, a large electrolytic capacitor at **Y** and a push to make (reset) switch connected between R1 and the negative line. In the table below are some suggestions for **X** and **Y**. You will have to try these for yourself as each electrolytic capacitor and variable resistor can vary considerably from the values marked on the can. Try different combinations and time them with a stop watch until you have the time range you require.

X	Y	Time in seconds
10 k variable	1000 μF	up to 5 s
10 k variable	2000 μF	up to 10 s
100 k variable	1000 μF	up to 40 s
100 k variable	2000 μF	up to 80 s

TIME DELAY CIRCUIT

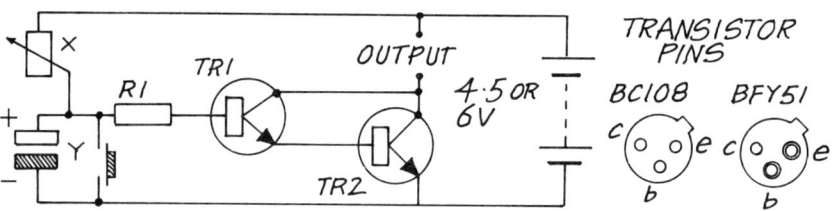

TIMER

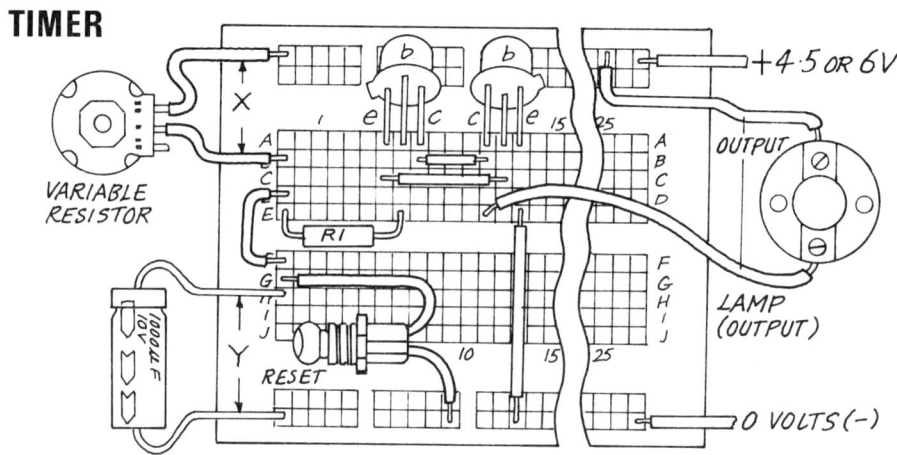

LIGHT SENSOR

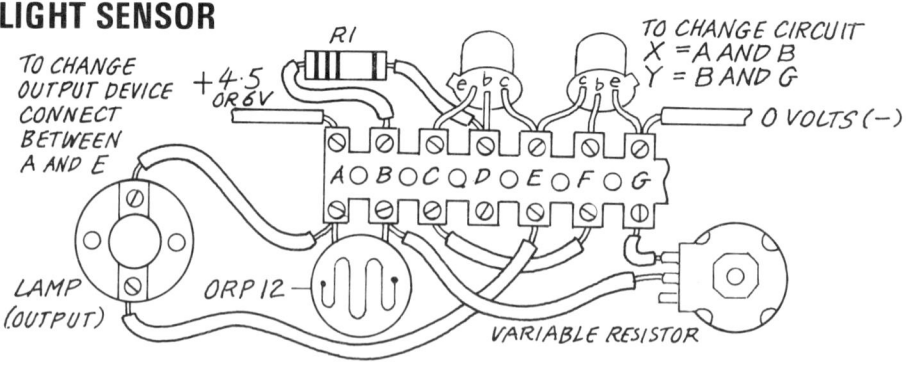

TO ADD A RESET SWITCH FOR A TIMER CONNECT BETWEEN B AND G

THE FLIP FLOP FLASHER UNIT

THE FLIP FLOP

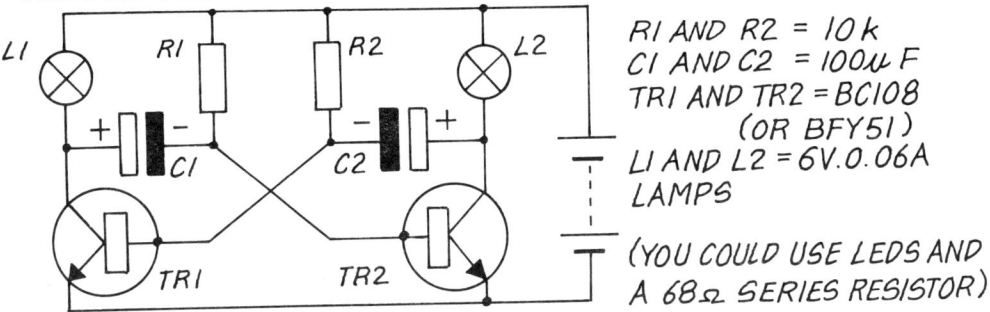

RI AND R2 = 10 k
CI AND C2 = 100µF
TRI AND TR2 = BC108
         (OR BFY51)
LI AND L2 = 6V.0.06A
LAMPS

(YOU COULD USE LEDS AND
A 68Ω SERIES RESISTOR)

The circuit is called an astable multivibrator and is two switches coupled with a time delay in such a way that they each control the other. Here is how it works. TR1, when on, charges C1. When C1 is full it discharges through R1 and the base of TR2 and so switches TR2 on. C2 by now has discharged itself and there is not enough current to keep TR1 on, so it switches off. Meanwhile TR2 charges C2. When C2 is full it discharges through R2 and the base of TR1 and so switches TR1 on again. By now C1 has discharged itself and there is not enough current to keep TR2 on, so it switches off. We are now back where we started and the cycle repeats itself.

When you set up this circuit you will see that each lamp is on and off alternately for about equal times with a flash rate of about 40 per minute. This flash rate can be changed very easily. Change R1 to 1 k. L1 now has a short on time with a long off time. L2 has a long on time with a short off time. If you replaced L2 with a 100 Ω resistor, you would have a single lamp unit which flashed with a short on time and long off. Alternatively L1 could be replaced with a 100 Ω resistor giving a single lamp unit flashing with a long on and short off. (Do not make R1 or R2 much less than 1 k or the transistors could be damaged.)

You can also alter the flash rate by changing the size of the electrolytic capacitors. Change R1 back to 10 k and C1 to 10 µF. The flash rate is the same as when R1 was 1 k. This is because the flash rate depends on C1 × R1 and C2 × R2. Now change R2 to 1 k. The flash rate has roughly equal on and off times. However the flash rate is much faster. How else could you have achieved this? Try using a buzzer instead of one of the lamps.

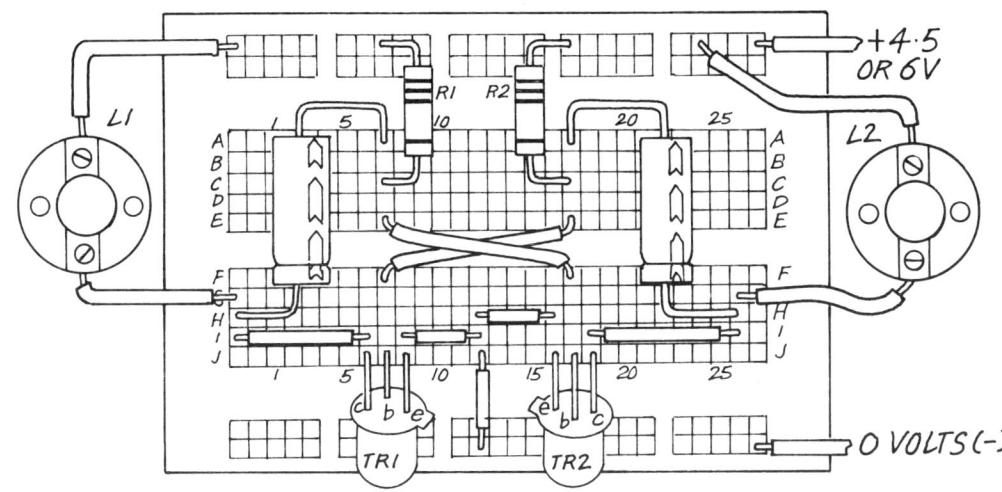

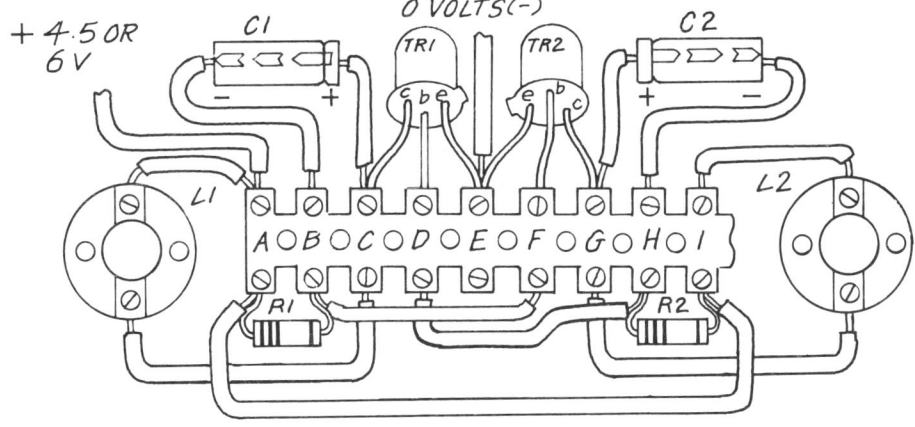

A FLIP FLOP BUZZER   THE LOUDSPEAKER

A FLIP FLOP BUZZER

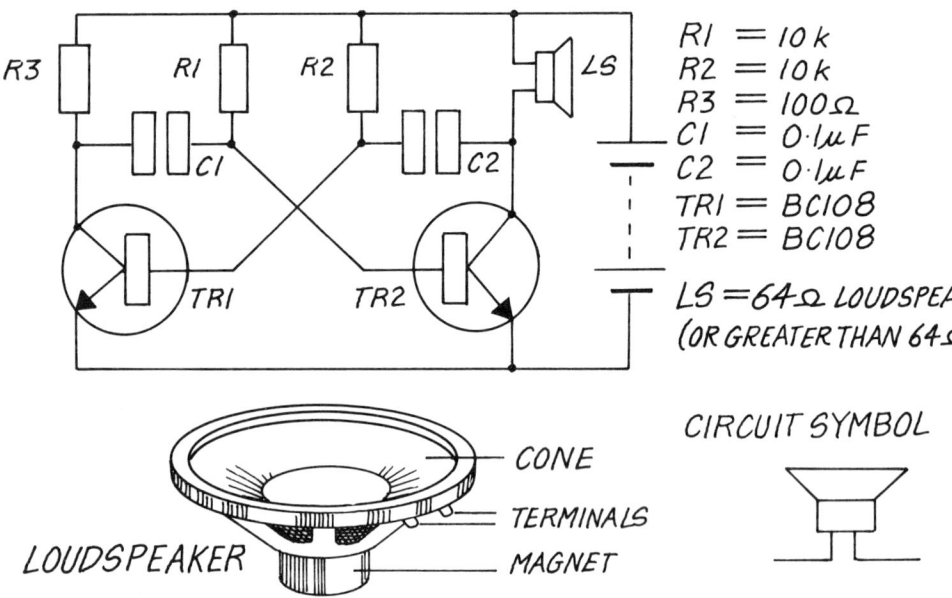

R1 = 10 k
R2 = 10 k
R3 = 100 Ω
C1 = 0·1 μF
C2 = 0·1 μF
TR1 = BC108
TR2 = BC108

LS = 64 Ω LOUDSPEAKER
(OR GREATER THAN 64 Ω)

CIRCUIT SYMBOL

CONE
TERMINALS
MAGNET

LOUDSPEAKER

We have a new component to introduce called a loudspeaker. It is an electromagnetic device which changes alternating currents (a.c.) at audio frequency (the same vibration rate as sound) into sound. (Remember all sound is made by something vibrating – the faster the vibration, the higher the note.)

THE CIRCUIT

We described the action of the astable multivibrator (flip flop) in Electronics 16. The main differences in this circuit are the size and type of capacitors, replacing L1 with a 100 Ω resistor (R3) and L2 with a 64 Ω loudspeaker. On switching on you should hear a note from the speaker. This is because the switching rate between TR1 and TR2 is at the same vibration rate (frequency) as sound. This electrical vibration makes the speaker cone vibrate

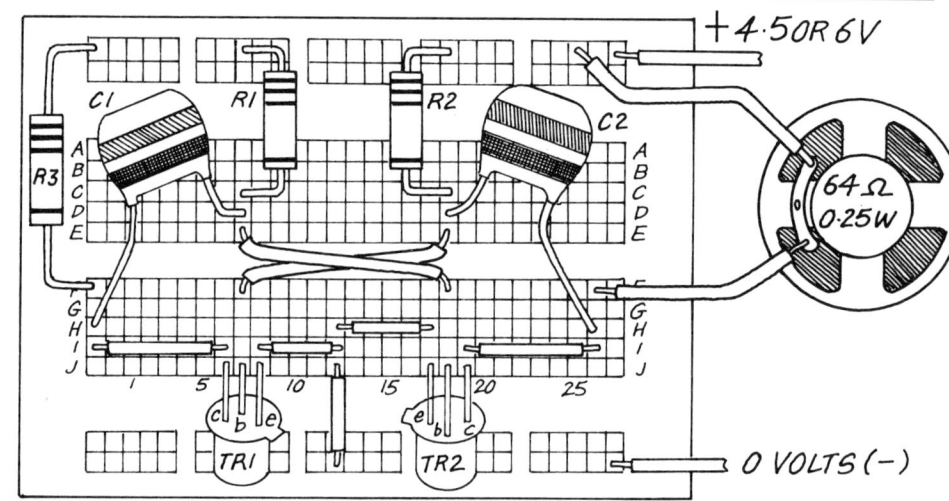

making a sound of the same frequency. To change the note make C1 = 0.01 μF: this gives a higher note (faster switching). Change C2 to 0.01 μF: this gives a still higher note. Change R1 to 1 k: this gives a yet higher note. Change R2 to 1 k: no sound at all! Well there is, but you would need to be a bat to hear it! Change everything back. You can also try putting an ORP12 or wet sensor or a 10 k variable (or preset) resistor in series with R2.

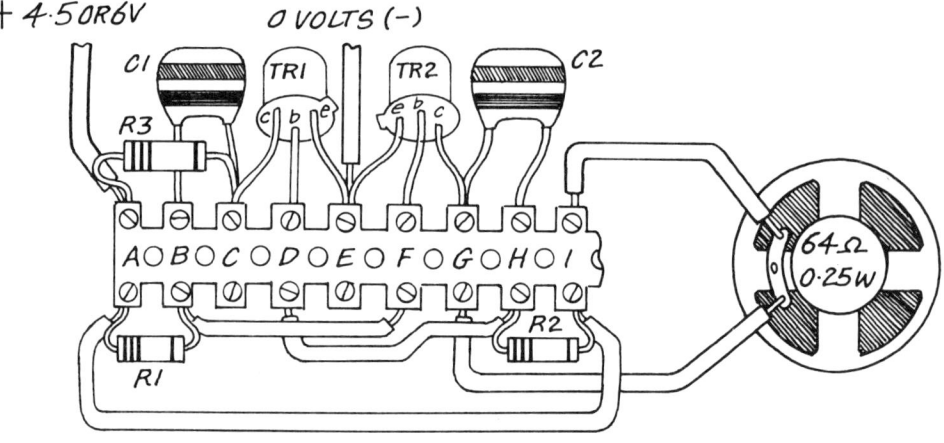

SOLDERING

THE ELECTRICAL SOLDERING IRON

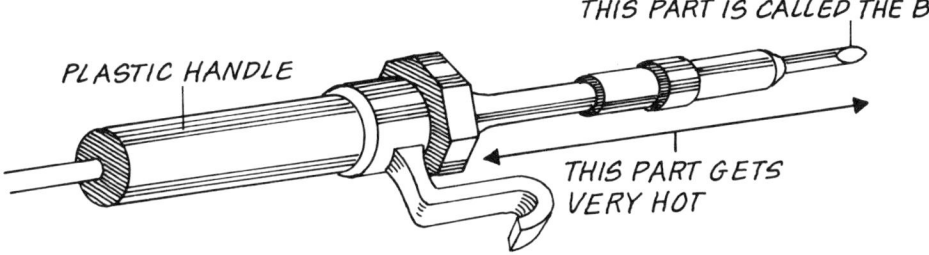

This is the best type of soldering iron to use for electronics. It is rated at about 15 or 17 watts. You should also have a proper stand to rest the iron on safely when not in use. The other main requirement is high grade 60/40 tin/lead solder 22SWG containing its own flux.

TINNING THE IRON

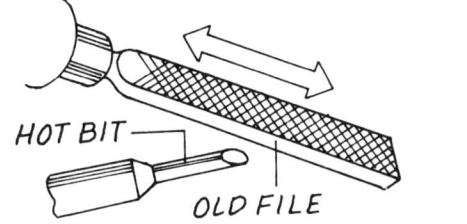
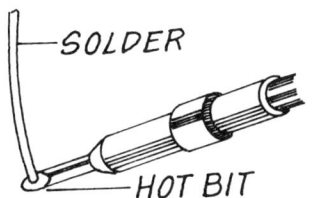

When the iron is hot, clean the end with an old file. Take care not to touch any of the metal part of the soldering iron – it's very hot! Touch some solder onto the cleaned bit so the tip becomes coated with a silvery layer of solder. Your iron is now tinned.

TINNING THE WIRE

Remove about 5–10 mm of the insulation from the end of the wire with wire strippers. Tin the wire by holding the bare end onto the hot bit and then touching the solder onto the wire. Take the solder away as soon as the wire is coated. It is easier to do this with the help of a friend as there are times when you seem to need at least three hands! If you are using stranded wire, twist the strands tightly together before tinning.

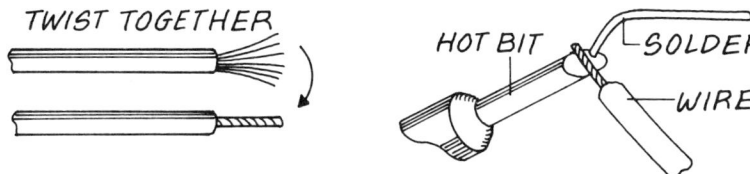

SOLDERING THE WIRE ONTO TAGS

Tin the metal solder tag by holding the hot bit against the tag. Wait about three seconds then touch the solder onto the tag near the bit. The solder should melt and coat the tag. Ask a friend to hold the tinned wire against the tag. Take care not to touch them with the soldering iron or to drip hot solder on them! Touch the hot bit onto the tag. Wait three seconds then touch on the solder. Do not put on too much solder. Take the solder away, then the iron. Your friend must hold the wire very still until the solder has set (normally 10 to 20 seconds).

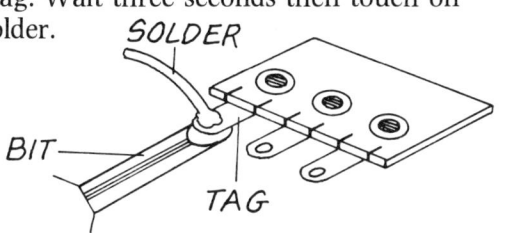

SOLDERING WIRES ONTO TRANSISTOR LEADS

Prepare the extension wires as before. Hold the transistor lead with pliers as shown below whenever applying the hot iron. This prevents heat damaging the insides of the transistor. Tin the transistor lead. Then solder on the extension wire as above. Repeat for each lead in turn.

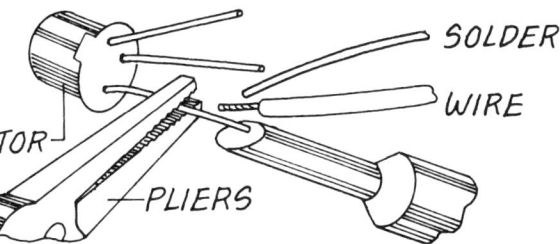

USING MATERIALS

Clockwise from left.   Welder wearing safety equipment, joining metal; kitchen equipment made from plastics; different types of wood: oak, mahogany, walnut, teak; some common adhesives; leisure boat made from glass reinforced plastic, GRP.   **How were the kitchen items made?**

SAFETY   SOME REMINDERS

YOURSELF

- Are you wearing a suitable apron or overall? **YOU SHOULD.**
- Are you wearing strong shoes? **YOU SHOULD.**
- Does your tie need tucking in?
- Are your sleeves flapping about dangerously?
- Does your hair need tying back?
- Don't get them caught in a machine. **MAKE THEM SAFE!**
- Should you be wearing goggles?
- Should you be wearing gloves?
- Should you be wearing a face mask?

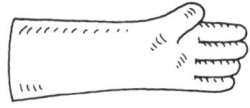 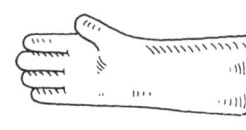 

YOUR WORK

Your work must always be held tightly before you try to use any tools.

Have you been shown how to use the tool you are about to use correctly? If not, ask your teacher to show you.

Have you left tools lying dangerously on top of your bench?

Clear all tools from your bench as soon as you are no longer using them.

THE MACHINERY

Have you been shown how to use the machine correctly?

Do you need permission to use this machine? If so, have you got that permission?

- You should be the *only* person in control of the machine.
- The safety guards *must* be in position.
- You *must* be wearing safety goggles.
- Your eyes are delicate and very, very precious.

GENERAL POINTS

If you are not sure, ask your teacher.

If a machine or tool is faulty, tell your teacher at once.

If you hurt yourself in any way, tell your teacher where and how.

If you see anything which might cause an accident, report it right away.

Read all labels and instructions carefully. When you have done that read them again.

THE TOOLS USED

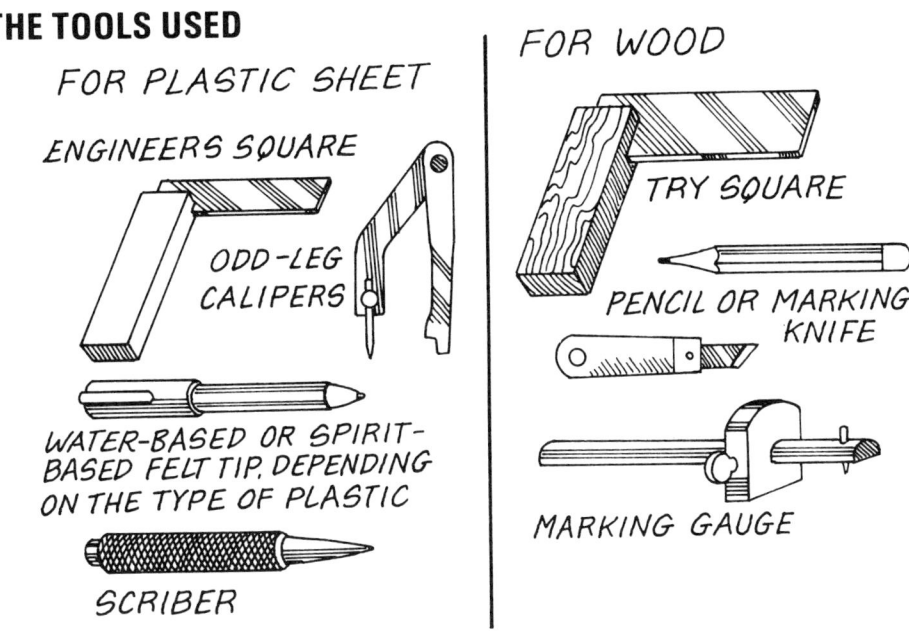

FOR PLASTIC SHEET

ENGINEERS SQUARE

ODD-LEG CALIPERS

WATER-BASED OR SPIRIT-BASED FELT TIP, DEPENDING ON THE TYPE OF PLASTIC

SCRIBER

FOR WOOD

TRY SQUARE

PENCIL OR MARKING KNIFE

MARKING GAUGE

USING A TEMPLATE

For awkward or irregular shapes, and for repeating the same shape, use a **template**. A template is made by cutting the required shape out of card. Place the shape on the material to be marked. Draw round the shape carefully with an appropriate marker. Note that we avoid waste by putting the template as close to the corner as possible.

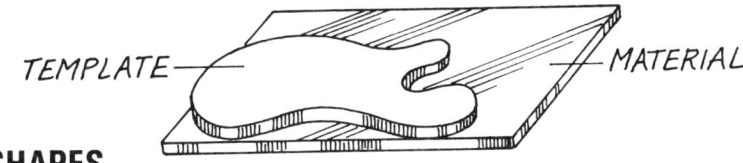

TEMPLATE — — MATERIAL

REGULAR SHAPES

Most of our work depends on straight lines and right-angles, so it is very important to be as accurate as possible in marking and checking them.

TESTING FOR A TRUE EDGE

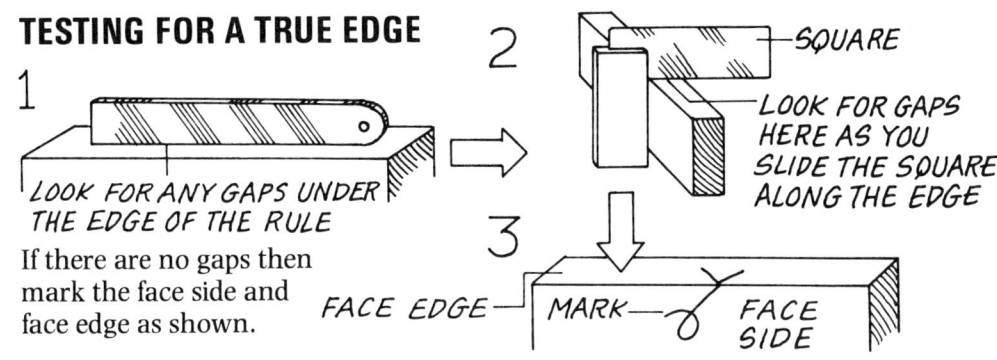

1   LOOK FOR ANY GAPS UNDER THE EDGE OF THE RULE

If there are no gaps then mark the face side and face edge as shown.

2   SQUARE

LOOK FOR GAPS HERE AS YOU SLIDE THE SQUARE ALONG THE EDGE

3

FACE EDGE — MARK   FACE SIDE

MARKING A SQUARE END

When you cut the material, cut on the *waste* side of the line.

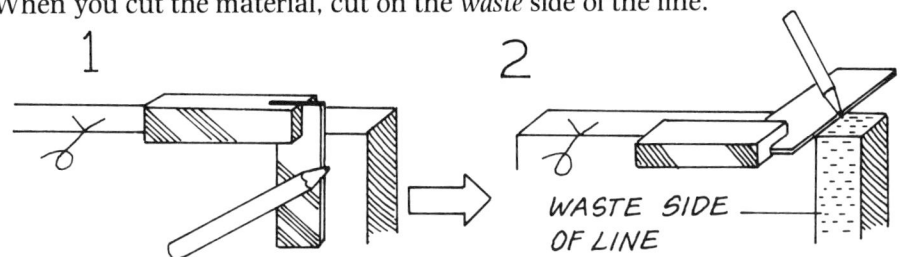

1   2   WASTE SIDE OF LINE

MARKING PARALLEL TO A TRUE EDGE

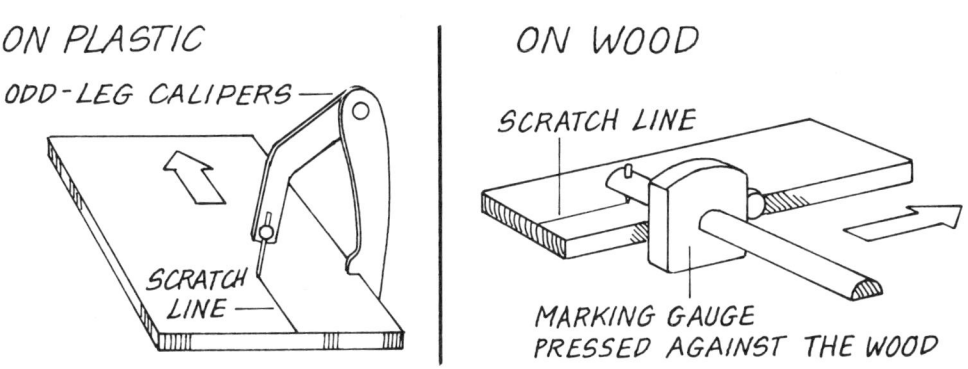

ON PLASTIC

ODD-LEG CALIPERS

SCRATCH LINE

ON WOOD

SCRATCH LINE

MARKING GAUGE PRESSED AGAINST THE WOOD

WHAT ARE PLASTICS?

Plastics are a group of substances that have long chain-like molecules called polymers. Although early plastics were natural or near-natural substances, most are now man-made from crude oil, coal and natural gas.

There are two main types of plastics, thermosetting and thermoplastic.

Thermosetting plastics are usually used in the workshop in the form of resins used with glass fibres – glass reinforced plastic (GRP) – or as **adhesives**. There are two common ones, polyester resin and epoxy resin. They come in two parts, the resin and the hardener, and must be used exactly as stated in the instructions given on (or with) the container.

EPOXY RESIN

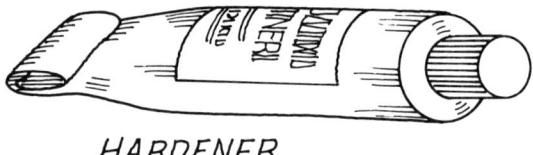

HARDENER

Thermoplastics are ones which soften when heated and harden again on cooling. The heat softening and cold hardening can be repeated many times. There are two thermoplastics commonly used in the workshop, acrylic (perspex) and polystyrene. Acrylic softens at about 180°C, when it is easily moulded. It becomes rigid again on cooling. It comes mainly in sheets of various thicknesses, though rods and tubes are available. It comes either as transparent, both clear and coloured, or opaque, both white and coloured.

Polystyrene comes in two very different forms, either as a foamed plastic as used in packaging and ceiling tiles, or as a rigid sheet. The rigid sheet softens at about 90°C and is often used in vacuum forming. Clear sheets about 2 mm thick are used for secondary glazing.

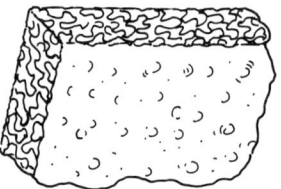 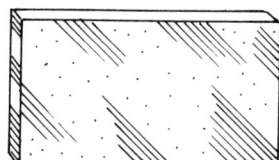

POLYSTYRENE CEILING TILE   RIGID POLYSTYRENE SHEET

Other common thermoplastics are polythene (high and low density), PVC (polyvinyl chloride) and nylon. Nylon may be used in the form of gear wheels, door catches, hinges and fishing line. These uses give some idea of its toughness. You will have used PVC as records. It is also used for guttering, drain pipes and corrugated roofing sheet. Softened PVC is used as a substitute for leather in coats and shoes. Polythene comes in the form of bags, buckets and washing-up bowls.

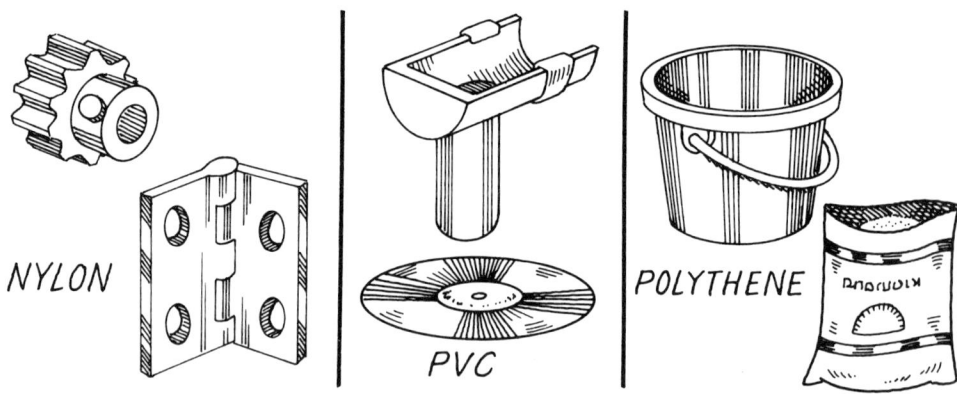

NYLON   PVC   POLYTHENE

Many packing containers are made from these materials so you may well use these plastics in using such free resources!

CUTTING

Plastic sheet (acrylic or polystyrene) can be cut to size by scoring and snapping. This is most useful for larger pieces of plastic and for straight lines.

SCORING

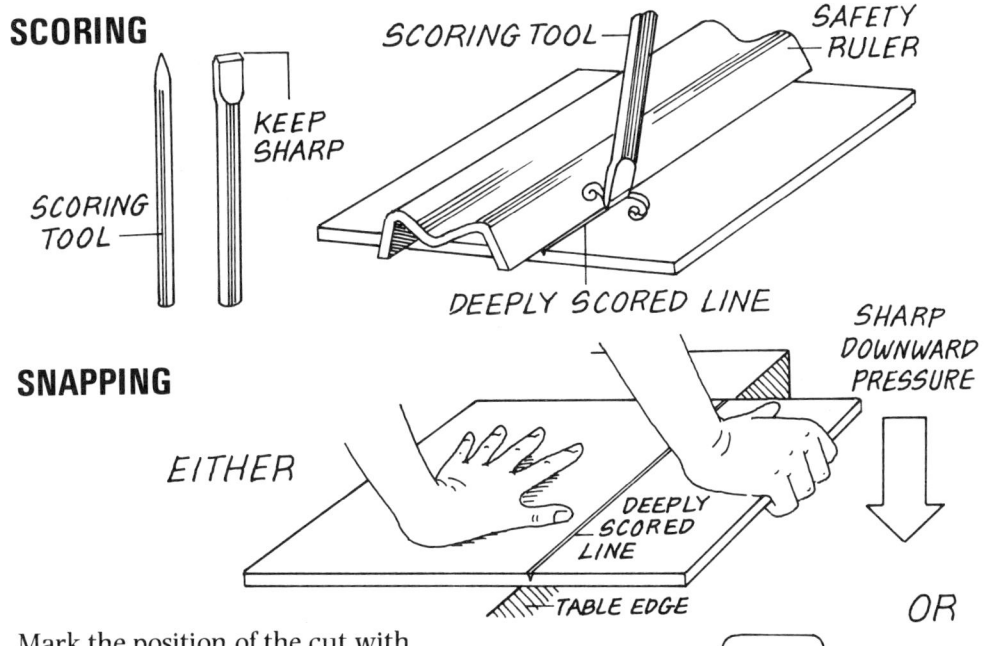

SNAPPING

Mark the position of the cut with a suitable marker. Using a safety ruler and a scoring tool, score a deep groove right from one edge to the other. (Important if the plastic is to break cleanly.) Position the scored line along the edge of the table (cramp if necessary) and snap the plastic with a sharp downward pressure.

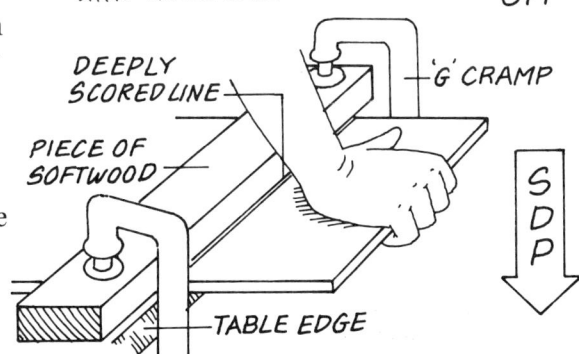

SAWING

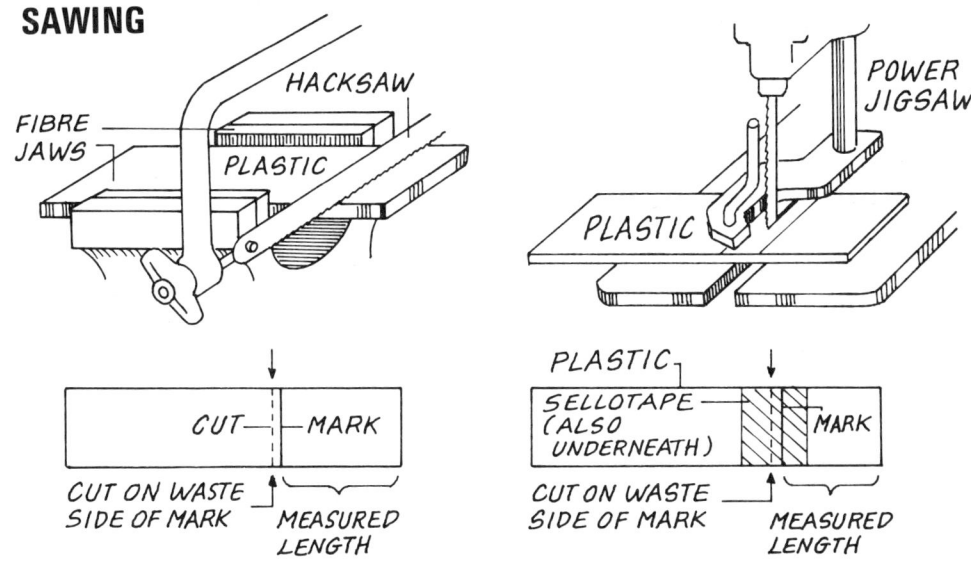

Short lengths can be cut using a hacksaw. Use fibre (soft) jaws in your vice and cut on the waste side of your line. Then you can file accurately back to the line as below (or use a sanding wheel if allowed). A power jigsaw (with the correct blade) can also be used, particularly for curves and irregular shapes. Remember to put sellotape above and below the cut to stop the plastic rewelding itself together or breaking the blade.

FILING

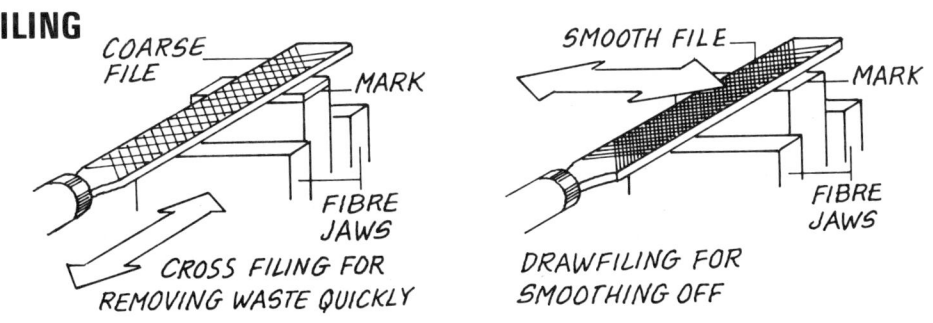

FORMING – FOLDING

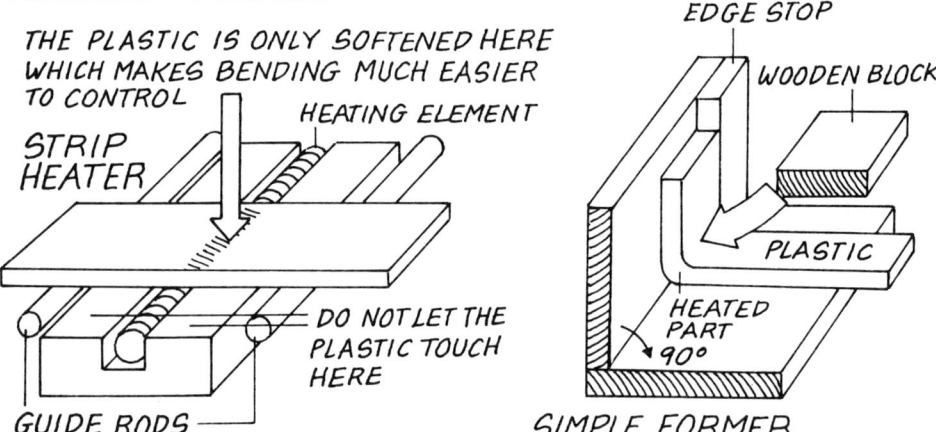

THE PLASTIC IS ONLY SOFTENED HERE WHICH MAKES BENDING MUCH EASIER TO CONTROL

STRIP HEATER

HEATING ELEMENT

DO NOT LET THE PLASTIC TOUCH HERE

GUIDE RODS

EDGE STOP

WOODEN BLOCK

PLASTIC

HEATED PART 90°

SIMPLE FORMER

To fold a plastic strip (acrylic or polystyrene) into a right-angle use the stripheater to heat the plastic along the line where the fold is required. Keep turning the plastic over to avoid overheating. (Remember acrylic softens at 180 °C, polystyrene at 90 °C.) When the plastic is soft enough to bend easily, position it in the simple former and press a wooden block into the fold until the plastic becomes rigid. (Wrapping the wood in cotton fabric will prevent grain marking.)

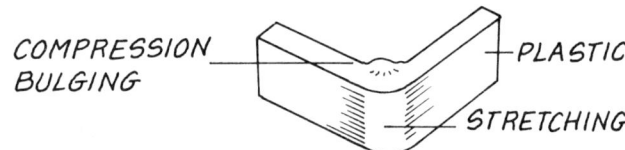

COMPRESSION BULGING — PLASTIC

— STRETCHING

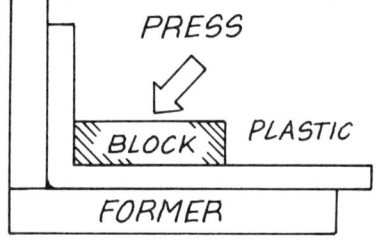

PRESS

BLOCK   PLASTIC

FORMER

Because the inside surface of the fold has been compressed and the outside stretched (tension), the edge of the fold bulges out as shown above. These bulges need to be filed down level before you can join them to a flat surface.

SIMPLE PRESS MOULDS

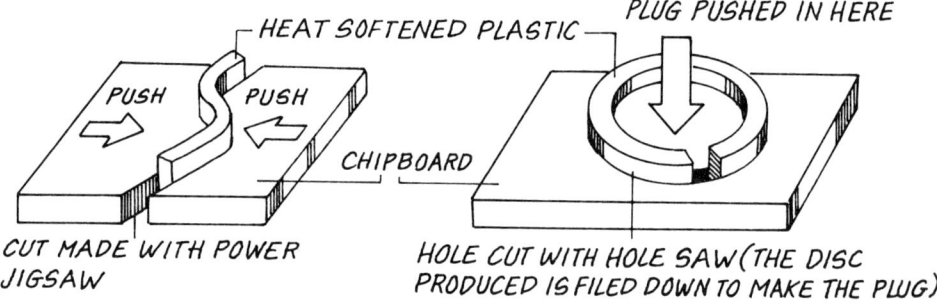

HEAT SOFTENED PLASTIC

PUSH   PUSH

CUT MADE WITH POWER JIGSAW

PLUG PUSHED IN HERE

CHIPBOARD

HOLE CUT WITH HOLE SAW (THE DISC PRODUCED IS FILED DOWN TO MAKE THE PLUG)

Here are two simple press moulds made from chipboard. The one on the left is made using the power jigsaw, the one on the right using a holesaw. The plastic is cut to the right length (use a strip of paper in the mould to measure). Put the plastic in the oven to soften (acrylic 180 °C and polystyrene 90 °C). When soft, it is placed in the mould and pressed until the plastic becomes rigid. (Make sure that the bottom edge is flat.)

PLUG AND YOKE MOULD

The plug should be slightly tapered. The diameter of the yoke must allow for a thickness of plastic all the way round. The plastic is softened and then the yoke presses it over the plug to form a plastic dome.

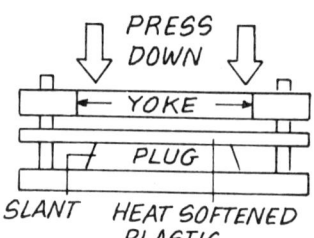

PRESS DOWN

YOKE

PLUG

SLANT   HEAT SOFTENED PLASTIC

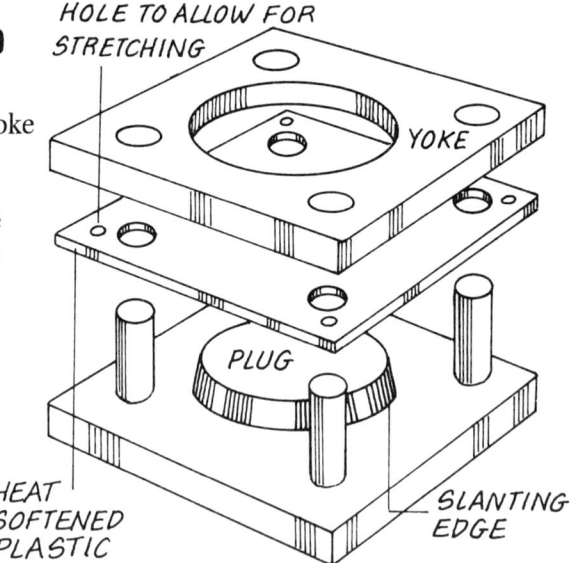

HOLE TO ALLOW FOR STRETCHING

YOKE

PLUG

HEAT SOFTENED PLASTIC

SLANTING EDGE

PLASTIC MEMORY

First make a simple mould as shown on the right. Make the shape you wish to use out of wire and fix it in the mould with double-sided sellotape. Cut the acrylic to size and heat in an oven at 180°C until soft. Place the acrylic in the mould and then press the topboard down on top using a G cramp or vice. Leave it to cool then remove the acrylic from the mould. Your acrylic should look something like this.

File the surface of the acrylic away almost to the bottom of the groove left by the wire. Try to keep your filing even.

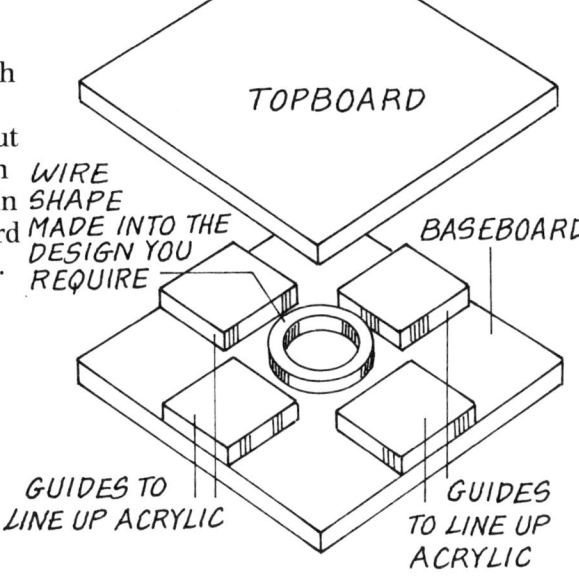

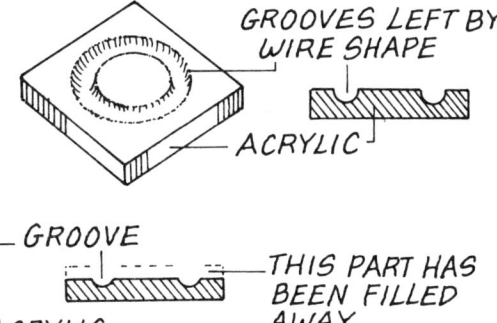

Put the acrylic back into the oven and reheat to 180°C. Take it out and leave to cool, by which time your acrylic should have the design as a shiny, raised section against the lower matt surface made by filing.

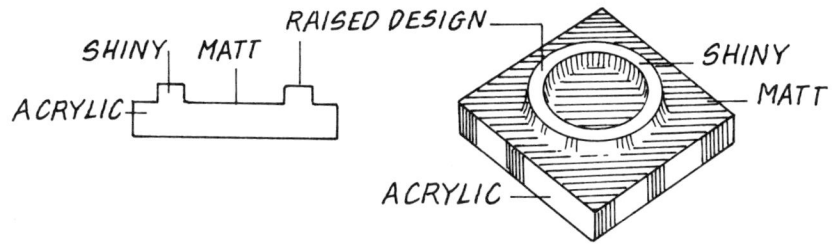

The plastic that was squashed by the wire has expanded to its original size, whilst the rest of the surface has been filed away.

DRILLING

The drill bits used to drill plastic – especially the large ones – should have been re-ground to a much shallower angle than normal otherwise the plastic tends to shatter. Make sure you have the right type of drill and cramp your work firmly when drilling plastic.

FINISHING PLASTIC

Remove completely any saw marks by filing (hold your plastic in fibre jaws). When all the saw marks have been removed, you can smooth your plastic further by drawfiling with a fine file or by using a metal scraper.

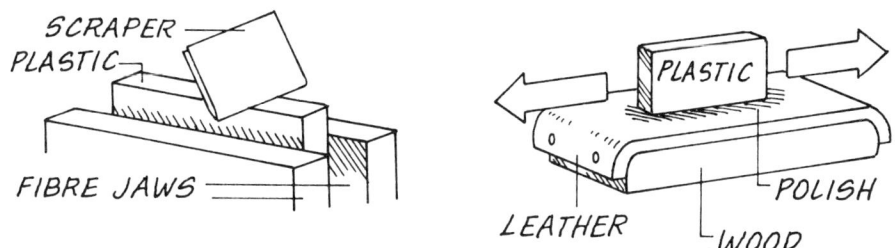

Polish your plastic by rubbing with fine grades of 'wet and dry' paper used wet. Finish off with perspex or metal polish (or buffing machine if allowed).

NATURAL WOOD

Trees are felled and cut into thick planks ready for seasoning. Seasoning is a drying process either natural (air dried) or using a special kiln. Wood for indoor work should dry until its **moisture content** is 10 %.

There are two main types of wood, called hardwood and softwood.

BROADLEAVED TREE (DECIDUOUS)

CONIFER (CONE-BEARING)

Hardwood is the name given to all woods coming from broad-leaved trees, i.e. trees which shed their leaves in winter. Softwood is wood which comes from conifers, i.e. cone-bearing trees. Common hardwoods are oak, beech, ash, mahogany, teak, sapele and balse (yes, balsa!). Common softwoods are pine (or red deal/redwood), cedar, spruce. Scots pine and yew. Wood being a natural product varies in quality quite a lot. It has an attractive surface due to its grain and is pleasant to the touch. It also suffers from some disadvantages. Insects and fungi will attack it unless it is protected.

Moisture affects it causing it to swell when wet and shrink when dry. It also tends to warp and twist and to split along the grain; wood is strong in one direction and not in the other. This is the reason for wood being processed into various manufactured boards.

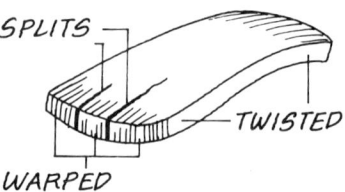

SPLITS

TWISTED

WARPED

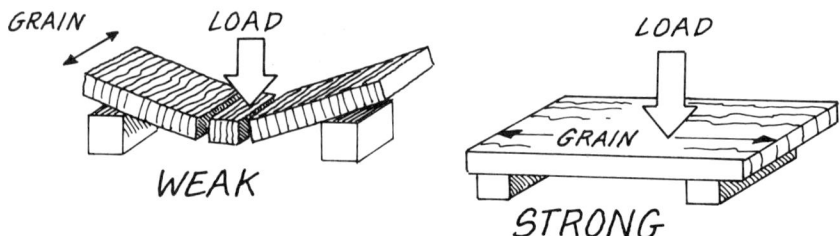

GRAIN   LOAD

WEAK

LOAD

GRAIN

STRONG

PROCESSED WOOD

Natural wood is made into veneer (thin sheets), wood chips or pulp. From these, various kinds of board are manufactured. They are equally strong in all directions.

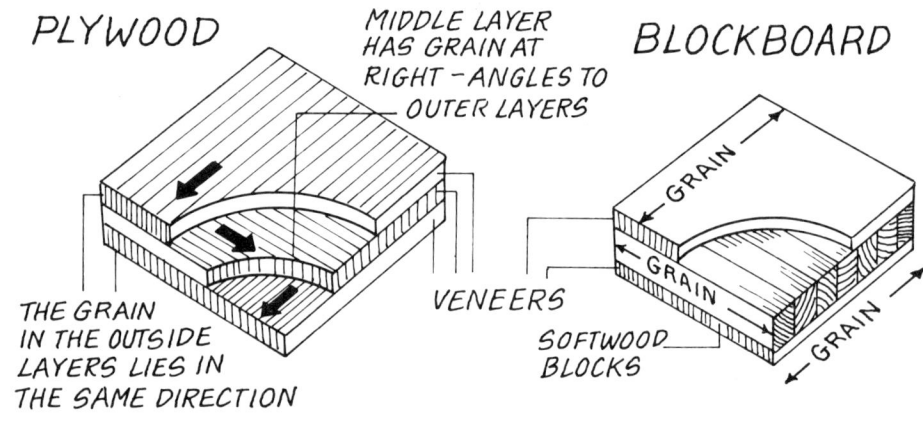

PLYWOOD

MIDDLE LAYER HAS GRAIN AT RIGHT-ANGLES TO OUTER LAYERS

BLOCKBOARD

THE GRAIN IN THE OUTSIDE LAYERS LIES IN THE SAME DIRECTION

VENEERS

SOFTWOOD BLOCKS

GRAIN

CHIPBOARD

WOODCHIPS AND GLUE BONDED AND PRESSED INTO A FLAT SHEET

HARDBOARD

WOOD PULP AND GLUE WHICH HAS BEEN PRESSED INTO A SHEET

SMOOTH SIDE

ROUGH SIDE

They do not warp, twist or split as easily as natural wood. Chipboard often has veneer or plastic glued as a finish to the outer surfaces. Hardboard has a rough side and a smooth side. Plywood can have more layers – there is always an odd number. Marine ply can be used outdoors.

SAWING WOOD

If we are making a straight cut in wood the best saw to use is a tenon saw. This has a piece of metal down the back edge of the saw to keep the blade straight. When you have marked out your wood, hold it in a vice, cramp or use a bench hook cramped in the vice as shown. Use the thumb of your non-sawing hand to guide the blade.

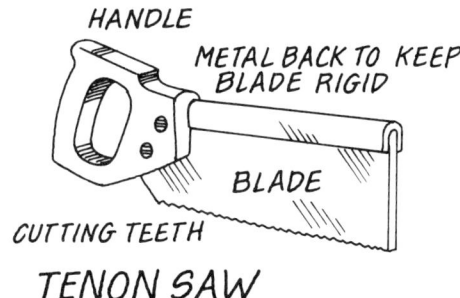

HANDLE
METAL BACK TO KEEP BLADE RIGID
BLADE
CUTTING TEETH
TENON SAW

WOOD MARKED AND MEASURED
BENCH HOOK
MARK
VICE
BENCH

Always start with two upward pulls of the saw before trying any push strokes. Use the full length of the blade. When you are almost through the wood, slow down and support the wood to avoid breaking off a splinter and spoiling the cut. Always saw your wood through the direction in which it is thinnest. Can you see why?

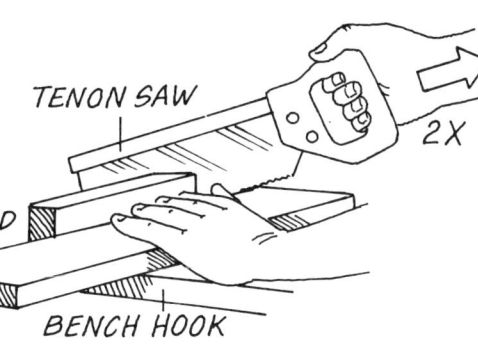

TENON SAW
2X
WOOD
BENCH HOOK

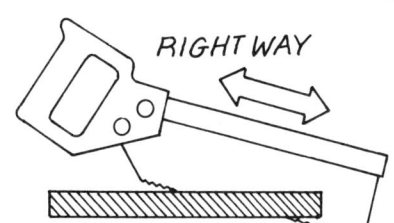

RIGHT WAY

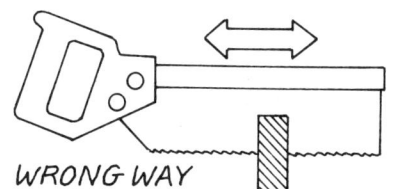

WRONG WAY

For curves and non-straight cuts, use a coping saw or a power jigsaw. (Remember that a coping saw cuts on the pull stroke while a tenon saw cuts on the push.)

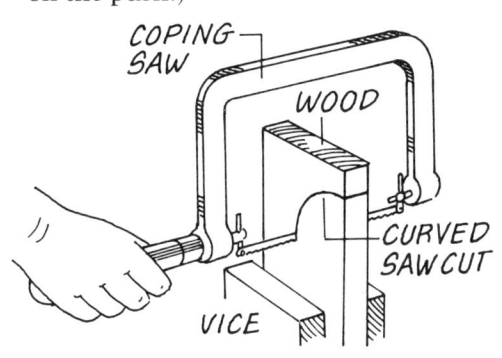

COPING SAW
WOOD
CURVED SAW CUT
VICE

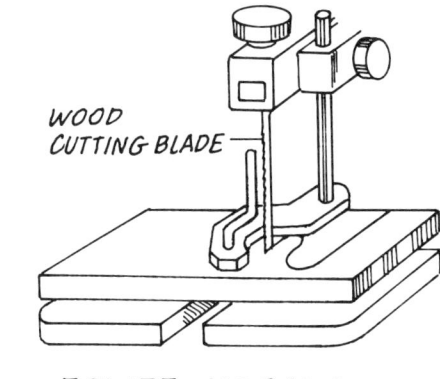

WOOD CUTTING BLADE
POWER JIGSAW

SHAPING WOOD

Here are a few of the tools with which wood can be shaped.

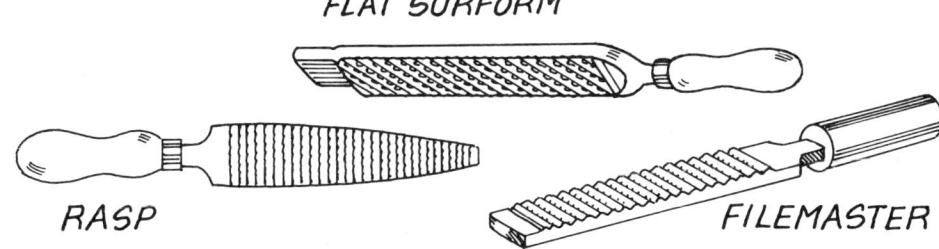

FLAT SURFORM
RASP
FILEMASTER

When shaping wood with these tools always have your work held in a vice or cramp. Hold the handle of the tool with your main hand then grip the other end of the tool with your other hand. This helps to make it easier to control and to give more pressure. All the tools shown above cut on the push stroke. They all leave the wood surface rather rough so it needs careful finishing afterwards.

DRILLING WOOD   USING CHISELS

DRILLS

The most common type of drill bit is the twist drill and can be used either in power drills or in hand-driven drills. The drill bit is gripped by a device called a 'chuck'.

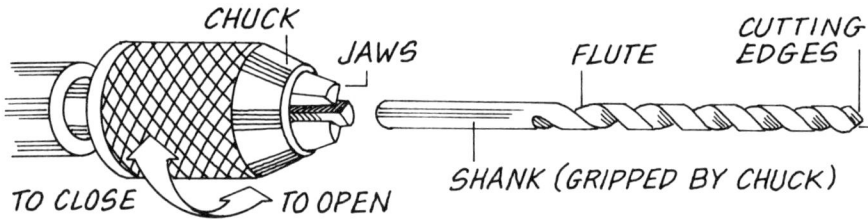

CHUCK JAWS FLUTE CUTTING EDGES
TO CLOSE TO OPEN SHANK (GRIPPED BY CHUCK)

The chuck shown here is one from a hand drill. The ones used on power drills have a toothed edge so they can be tightened with a chuck key.

DRILLING

Your work must be held in a cramp or vice. You should always cramp some scrap wood behind or under your work. This protects the back of your work from splintering when the drill comes through. Always protect the surface you are drilling on – bench top or drill table – with scrap wood. If you are drilling a deep hole take the bit out of the hole a few times to stop the drill clogging and getting too hot.

OTHER BITS

Here are two other types of bit. They are for use on power drills only. They are unsuitable for hand drills.

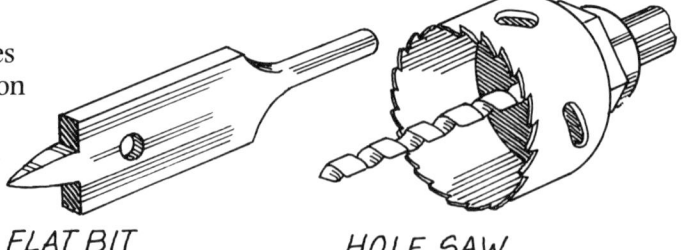

FLAT BIT   HOLE SAW

CHISELS

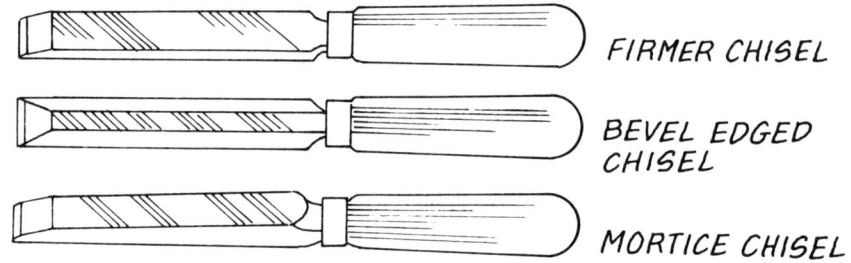

FIRMER CHISEL

BEVEL EDGED CHISEL

MORTICE CHISEL

Chisels split the wood along its grain very easily. A split, once started, usually goes further than we would like and, because it follows the grain, it doesn't always go in the right direction. For these reasons, it is important to make a limit to the wood we can remove, by making a saw cut where we want to stop.

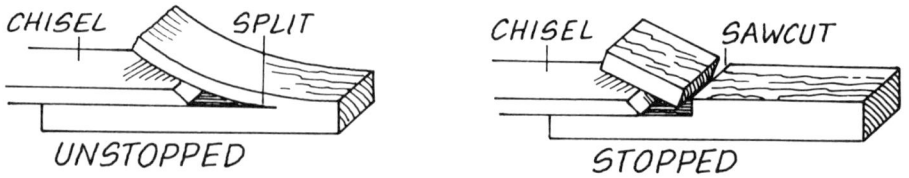

CHISEL SPLIT   CHISEL SAWCUT
UNSTOPPED   STOPPED

It is also easier to control the chisel cutting across the grain rather than along.

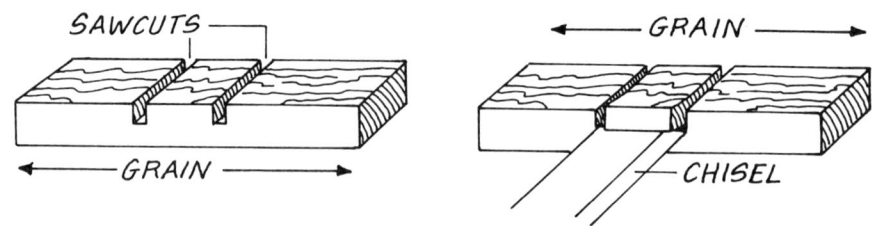

SAWCUTS   GRAIN
GRAIN   CHISEL

When using a chisel always have your work firmly cramped or in a vice. Keep *both* hands behind the cutting edge and always cut away from yourself. Always make a saw cut first and try to chisel across the grain. If you need to strike the handle of the chisel, always use a wooden mallet, not a hammer.

JOINTS

Because wood cannot be folded into a right-angle as readily as plastic or metal, corners in wood have to be cut and joined. Here are some of the simple ways that wood is joined.

BUTT JOINT

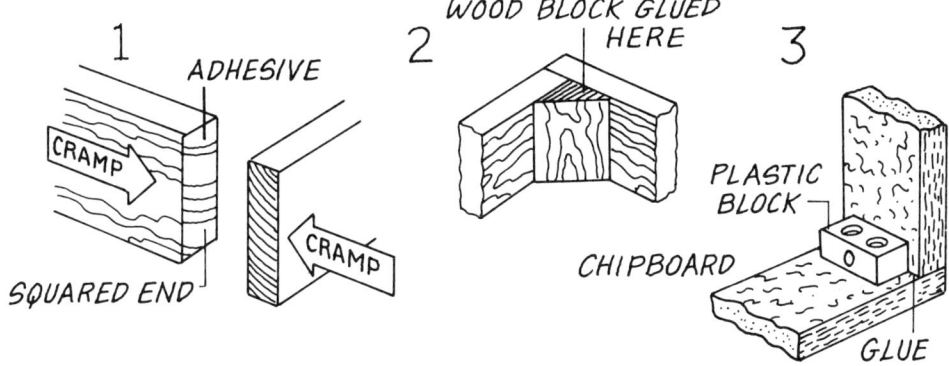

The butt joint (1) is the simplest joint. It is formed by gluing and cramping the squared end of one piece of wood against the side of the other. It is not a very strong joint as it depends mainly on the strength of the glue and the area glued. It can be strengthened by gluing in a corner block (2) or, in the case of chipboard, by using screw-on plastic corner blocks (3).

REBATE JOINT

This is a stronger joint than the butt joint. It is formed by gluing and cramping the squared-end of one piece into the specially cut channel (called a rebate) in the other.

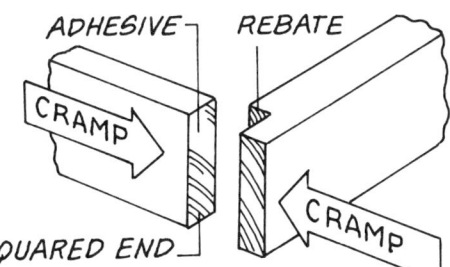

DOWEL JOINT

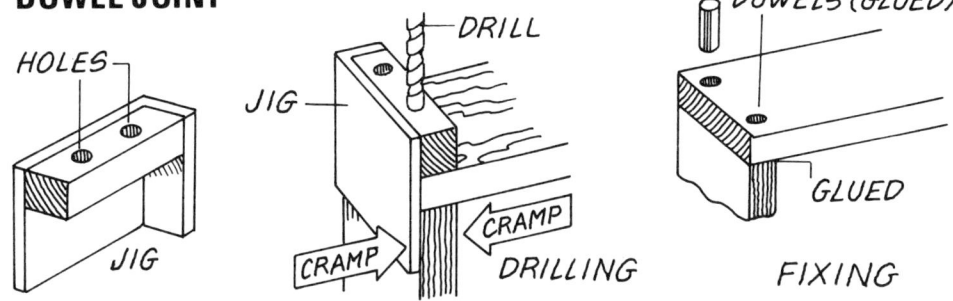

This is a stronger form of the butt joint. Holes are drilled through the overlapping piece into the end of the other. The pieces are glued together and then lengths of dowel (hardwood round rods) are glued into the holes. The holes are usually drilled using a jig to make sure the corner is square.

FINISHING WOOD

Wood must be made smooth before applying varnish or paint. The easiest way to do this is to use glasspaper. This comes in four grades: coarse (S2), medium (F2), fine (0) and flour (00). Start with coarse and do not use medium until you have removed all saw and other tool marks (or glue spots). Always rub in the direction of the grain. Work your way steadily through medium, then fine, then flour. Damp your work and leave to dry. Remove any roughness on drying with flour grade glasspaper and damp again. Repeat this until your wood dries smooth. Your wood is now ready for painting, oiling with linseed or teak oils or varnishing (sometimes the wood is dyed or stained first). Paint should be applied in three coats.

ADHESIVES   FIXINGS

ADHESIVES

The secret of a strong glued joint lies in choosing the right adhesive for the materials being joined, a large area of surface being glued and making sure that the surfaces being glued are clean. Here is a table to help you choose the right glue.

Adhesive	Description	Materials joined
PVA (e.g. Resin W)	Thick white liquid which dries colourless. It is *not* waterproof.	Wood to wood Fabric to wood Polystyrene foam to most materials
Synthetic Wood glue (e.g. Cascamite)	Powder which is mixed with water. *Waterproof* when set.	Wood to wood
Epoxy resin (e.g. Araldite)	A two-part adhesive. (The adhesive is in one tube, the hardener in the other. See Using Materials 3.)	Metal to metal Metal to wood Metal to acrylic Acrylic to wood Acrylic to acrylic
Tensol cement	Clear syrupy liquid with a nasty smell.	Acrylic to acrylic
Polystyrene cement	Clear liquid	*Rigid* polystyrene to *rigid* polystyrene
Contact or impact	These are rubber-based with a strong smell.	Fabric, rubber and plastic laminates to most materials
Hot glue gun	Solid sticks of glue of two types: for wood, or for metals and plastics.	It can be used for most materials but only on *small* areas

Remember most glues are not 'instant' and the two surfaces being joined will usually need cramping together until the glue has set.

FIXINGS

Fixing	Descriptions and limitations	Materials
PANEL PIN	This is for fixing thin sheets (e.g. 3-ply or hardboard) to a frame.	Sheet 'wood' to wood
OVAL NAIL ROUND WIRE	Oval nails do not hold as well as round nails but are less likely to split the wood. They can also be punched below the surface and so are less obvious.	Wood to wood
ROUND COUNTERSUNK RAISED (TYPES OF WOODSCREW)	Woodscrews make stronger joints than nails. (Do not screw into end-grain.) They are made of steel or brass. Their heads are slotted, either straight or crossed and are round, raised or counter-sunk. A hole is usually drilled before putting in the screw. SHANK SIZE OF SCREW NARROW PILOT HOLE COUNTERSUNK (IF NEEDED)	Wood to wood
CHEESEHEAD ALSO ROUND AND IN COUNTERSUNK	Self-tapping screws are made of hard steel. They make their own thread in the materials they join. They need a small hole drilling the same size as the screw core.	Metal to metal. Plastics that are not too brittle. Chipboard.
HEXAGONAL NUT AND BOLT (MANY OTHER TYPES)	There are many different kinds of nuts and bolts. They are used for joins that may have to be taken apart. They are often used with washers. PLAIN WASHER SPRING WASHER	Wood Metals Plastics

GLOSSARY

adhesive — sticky substance used to join surfaces together

airscrew — aircraft propeller

alignment — arranged in a line

amplification — increase of strength of a signal

amplifier — device which increases the strength of a signal (electrical message) in a circuit

assembly — group of connected parts

astable multivibrator — two coupled switches which each control the other

axis — line about which an object rotates

axle — shaft on (or with) which a device revolves

base — transistor control lead

bearings — the holes in which a shaft rotates

breadboard — circuit prototype board

buzzer — device which changes electrical energy to sound energy

capacitor — device for short-term storage of electricity

cathode — negative lead of a device

cellular — made of small cells or boxes

chassis — base frame of vehicle

collector — transistor lead forming the output of the controlled circuit

component — individual part of a whole circuit

compression — pressing together to reduce volume

conductor — material which allows the passage of electricity

corrugation — a wrinkle or fold

corrugated — folded into a series of ridges and grooves

crank — device for changing direction of movement

current alternating (a.c.) — current which rapidly changes direction from forwards to backwards to forwards etc.

current direct (d.c.) — current which stays fixed in one direction

Darlington pair — two transistors coupled to form a device for increasing current

device — something designed to do a particular job

diameter — straight line passing through the centre of a circle and connecting two points on the circumference which are opposite each other

diode — device that allows electrical current to flow in one direction only

displace — push out of the way

efficiency — useful work done by a machine compared with the energy put in

electromagnet — magnet made by the flow of electricity through a wire

emitter — transistor lead forming the input of the controlled circuit

evaluate — work out how well an activity has succeeded

exhaust — use up

filament — thin wire, resistant to melting, which glows white hot when an electrical current is passed through

flexible — easy to bend

friction — resistance as two bodies rub together

fulcrum — pivot point of lever

hopper — funnel

horizontal — level, parallel to horizon

inertia — the ability of an object to resist movement or change in movement

insulate — act as a barrier

insulator (electrical) — material which acts as a barrier to the passage of electricity

internal combustion engine — engine in which power is produced by burning a mixture of petrol, gas and air

light-dependent resistor (LDR) — device which decreases in resistance as the brightness of light increases

light-emitting diode (LED) — device which, if the current is in the right direction, changes electrical energy into light energy

linear — in line

magnetic field — space in which magnetic properties can be noticed

mechanism — system of moving parts

mock-up — rough model

moisture content — amount of liquid contained within timber

momentum — increase of force produced by the movement of an object

motion — movement

GLOSSARY

negative — side of a battery to which it is assumed a current will flow

oscillating — repeated backwards and forwards movement in a curve

parallel — lines which are always the same distance apart; in electronics, an arrangement of devices side by side sharing the available current

parallel (in) — sharing of available current between two or more devices

partitions — sections

pivot — short shaft on which something turns

positive — the side of a battery from which it is assumed a current will flow

potentiometer — device for varying electrical resistance

principle — basic truth in science

properties — main features

proportional — relating in size

prototype — first practical try-out of an idea

pulley — grooved wheel for a rope or belt to pass over

r.p.m. (revolutions per minute) — number of turns every minute

ratio — comparison of the number of times a big number can be divided by a smaller one

reciprocating — repeated backwards and forwards movement in a straight line

rectifier — device for converting alternating current into direct current

resistor — device which attempts to prevent the passage of electrical current

resistor (preset) — device, the resistance of which can be set to the required level

resistor (variable) — device, the resistance of which is variable from zero to a maximum stated value

revolution — turn

rigidity — stiffness, resistance to bending

rotary — circular movement about a fixed point

rotate — turn about a fixed point

series (in) — arrangement of devices one after the other in a chain

stabiliser — device to help prevent unwanted movement

static — not moving

synthetic — man-made, artificial

template — guide used for marking out shapes (useful when repeating them)

terminal block — a series of insulated connectors with a clamping screw at each end of every conductor

thermistor — heat-dependent resistor

traction — pulling power

transistor — electronic switch or amplifier

transmission — system of shafts and gears or pulleys by which the power is taken from the motor where it is produced to where it is needed

vacuum — completely empty space

vertical — upright

vibration — rapid movements to and fro

winch — see *windlass*

windlass — device for winding in a rope or chain

INDEX